# COLOURED QUADRANGLES

*To Mazurka and her Family*
*– my inspiring Lot*

OPUSCULA GRAECOLATINA

(Supplementa *Musei Tusculani*)

Edenda curavit Ivan Boserup

Vol. 24

# COLOURED QUADRANGLES

## A Guide to the Tenth Book of EUCLID's ELEMENTS

by
C. M. Taisbak

MUSEUM TUSCULANUM PRESS
COPENHAGEN 1982

© Museum Tusculanum Press 1982
Printed in Denmark by Special-Trykkeriet Viborg a-s
ISBN 87 8807 346 7
ISSN 0107-8089

ACKNOWLEDGEMENTS

We are indebted to Eivind Lorenzen, city architect of Copenhagen, for drawing the figures and laying out the cover. And to Kirsten Busch-Nielsen, undergraduate, and Peter Juul Trosborg, M. A., for reading our treatise at the proof stage. And to several teams of students who did their best to believe us whenever we changed our mind.

The costs of printing were met with a grant from the Faculty of Humanities in Copenhagen University.

# TABLE OF CONTENTS

|     | A Motivation: The Regular Pentagon |   |   |   | p. 9 |
|-----|---|---|---|---|---|
|     | XIII 11 | p. 9 | XIII 3 | p. 12 |
|     | XIII 9 | p. 11 | XIII 10 | p. 13 |
| 1   | Tools and Prerequisites |   |   |   | p. 16 |
| 1.0 | The Text |   |   |   | p. 16 |
| 1.1 | Conventions and Notations |   |   |   | p. 16 |
| 1.2 | The Premises. Our Thesis of Simplicity |   |   |   | p. 17 |
| 1.3 | The Second Book of the Elements |   |   |   | p. 18 |
|     | II 4-10 | p. 18 | I 47-48 | p. 19 |
|     | II 11 | p. 19 | I 43 | p. 19 |
|     | II 14 | p. 19 | I 44 | p. 20 |
|     | II def 1 | p. 19 |   |   |
| 1.4 | Theory of Ratio and Commensurability |   |   |   | p. 21 |
|     | X def 1 | p. 21 | V 16 | p. 24 |
|     | X 5 | p. 21 | V 22 | p. 24 |
|     | 200 def to X 5 | p. 22 | VI 22 | p. 24 |
|     | X 6 | p. 22 | VI 19 | p. 24 |
|     | X 11 | p. 22 | X 12 | p. 24 |
|     | VI 1 | p. 22 | X 13 | p. 24 |
|     | V 7 | p. 23 | X 15 | p. 25 |
|     | V 11 | p. 23 | X 16 | p. 25 |
|     | VI 16 | p. 24 | X 9 | p. 25 |
|     | VI 17 | p. 24 |   |   |

| | | | | | |
|---|---|---|---|---|---|
| 2 | Coloured Quadrangles | | | | p. 26 |
| 2.0 | Why Colours? An initiation into not translating | | | | p. 26 |
| 2.1 | Red Quadrangles and Red Line Segments | | | | p. 27 |
| | 201 | X deff 3 & 4 | p. 27 | 205 X 19 | p. 29 |
| | 202 | X def 4 | p. 28 | 206 X 20 | p. 30 |
| | 203 | X def 3 | p. 28 | 207 X 21 | p. 30 |
| | 204 | | p. 29 | | |
| 2.2 | Amber Quadrangles and Amber Line Segments | | | | p. 30 |
| | 208 | X 21 | p. 30 | 212 X 24 | p. 33 |
| | 209 | X 21 | p. 31 | 213 X 53/54 | p. 33 |
| | 210 | X 22 | p. 31 | 214 X 25 | p. 34 |
| | 211 | X 23 | p. 32 | 215 | p. 34 |
| | | | | 216 X deff 3 & 4 | p. 35 |
| 2.3 | Obscure Quadrangles. Sums and Differences of Line Segments | | | | p. 36 |
| | 217 | X 36 | p. 36 | 219 X 26 | p. 37 |
| | 218 | X 73 | p. 36 | | |
| 2.4 | Partition of Equal Quadrangles. Elliptic Application of Area | | | | 38 |
| | 220.1 | | p. 39 | 220.2 | p. 42 |
| | 221.1 | | p. 39 | 221.2 | p. 43 |
| | | X 17 & 18 | p. 41 | | |
| 2.5 | Six Classes of Sums-of-Two-Reds and Apotomes | | | | p. 43 |
| | 222 | X 17 & 18 | p. 43 | 225 X 42 & 79 | p. 47 |
| | 223 | X 60 & 97 | p. 43 | 223 bis | p. 48 |
| | 224 | X 48-53 & 85-90 | p. 44 | | |
| 2.6 | The Greater, the Lesser, and their Family | | | | p. 49 |
| | 226 | X 54, 91, 36, 73 | | | p. 49 |
| | 227 | X 55, 92, 37, 74 | | | p. 52 |
| | 228 | X 56, 93, 38, 75 | | | p. 52 |
| | 229 | X 57, 94, 39, 76 | | | p. 53 |
| | 230 | X 58, 95, 40, 77 | | | p. 54 |
| | 231 | X 59, 96, 41, 78 | | | p. 54 |
| | 229 | converse | p. 55 | 234 X 41/42 | p. 56 |
| | 232 | | p. 56 | 235 X 111 | p. 57 |
| | 233 | X 43-47, 80-84 | p. 56 | X 14 | p. 57 |

| | | |
|---|---|---|
| 2.7 | Postlude: Friendship between Sums and Apotomes | p. 58 |
| | X 112-113    p. 58              X 114 | p. 58 |
| 3 | By Way of Commentary | p. 62 |
| 3.1 | Genesis of the X'th book: The Regular Pentagon? | p. 62 |
| 3.2 | Crux Mathematicorum? Or the By-Heart-Way of Learning? | p. 65 |
| 3.3 | An unauthorized arithmetical interpretation | p. 66 |
| 3.4 | Appendix A. Some proofs | p. 70 |
| 3.5 | Appendix B. The word *dynamis* | p. 72 |
| | Bibliographical Epilogue | p. 77 |
| | Index of special terms | p. 78 |

I L L U S T R A T I O N S

| | | | | | | | | |
|---|---|---|---|---|---|---|---|---|
| Figure | 1 | p. 11 | Figure | 6 | p. 32 | Figure | 11 | p. 40 |
| | 2 | p. 14 | | 7 | p. 33 | | 12 | p. 42 |
| | 3 | p. 20 | | 8 | p. 33 | | 13 | p. 60 |
| | 4 | p. 23 | | 9 | p. 37 | | 14 | p. 60 |
| | 5 | p. 29 | | 10 | p. 40 | | 15 | p. 59 |

Diagram of important theorems    p. 50

A D V I C E

The arguments of this booklet should be visualized: if the relevant drawing happens to be overleaf, you will find it helpful to make your own informal sketch. The more so, if there happens to be no drawing.

A   M o t i v a t i o n :        T h e   R e g u l a r   P e n t a g o n

If the radius of a circle has the length $r$, and a regular pentagon be inscribed in that circle, then the side of the pentagon has the length
$$r/2 \cdot \sqrt{10 - 2\sqrt{5}},$$
and the diagonal of the pentagon has the length
$$r/2 \cdot \sqrt{10 + 2\sqrt{5}}.$$
This is what we expect to be told about the regular pentagon, and so did the Ancient Greeks:   In Euclid's Elements XIII 12 we learn that "if an equilateral triangle be inscribed in a circle, the square on the side of the triangle is triple of the square on the radius of the circle".
That is:  if the name of the radius be  r,  and that of the side be  s, we learn that          □ s = 3 □ r;
and in our modern algebraic idiom: if the length of the radius be  $r$, and that of the side be  $s$,  we learn that          $s = r\sqrt{3}$.

  Being so instructed about the regular triangle, the Greeks might also expect some precise comparison of the side of the pentagon with the radius or diameter of the circumscribed circle. But in Euclid's Elements they were *not* told these truths, at any rate not the whole truths:
In  XIII 11  we learn that

> "if an equilateral pentagon be inscribed in a circle
> which has its diameter *rhētḗ*, the side of the pentagon
> is *álogos*, the so-called *elássōn*."

On the face of it, this statement is very different from ours, as it involves no numerical constants, unless they be latent in the words that we hesitate to translate: *rhētē* "utterable", *álogos* "irrational", and *elássōn* "lesser", - seeing that they are obviously used in a very specific idiom. In this paper we shall initiate our reader into that idiom, thus enabling him to evaluate for himself the scope and limitation of the statement in XIII 11.


0.1 Some would argue that the Greeks were not able to put such complicated facts into words, and as they had no algebraic symbols, that was that! But this argument cannot hold, as you will realise when considering a regular decagon (figure 1), incsribed in a circle. (For the sake of clarity we do not draw *all* its sides and diagonals, but mark off the vertices on the circle, numbered 0, 1, ..., 9. The center of the circle is C.)

To evaluate the side and diagonal of the pentagon, we consider two right-angled triangles, 025 and 045, both having the diameter for their hypotenuse. Using I 47, the "Theorem of Pythagoras", we write the equations:

$$\square \underline{02} + \square \underline{25} = \square \underline{05} \qquad (1)$$
$$\square \underline{04} + \square \underline{45} = \square \underline{05} \qquad (2)$$

where 02 is the side and 04 the diagonal of the pentagon, and 45 is the side of the decagon; on the chord 25 we establish two points of intersection with diagonals of the decagon:

The diagonal 93 intersects 25 in A, and

the diagonal 94 intersects 25 in B, passing through the center C. Among other things, this lands us with some equilateral triangles, e.g.: B54, B2C, B92, - all of which, by the way, have their base angles twice their top angle.

*Coloured Quadrangles*

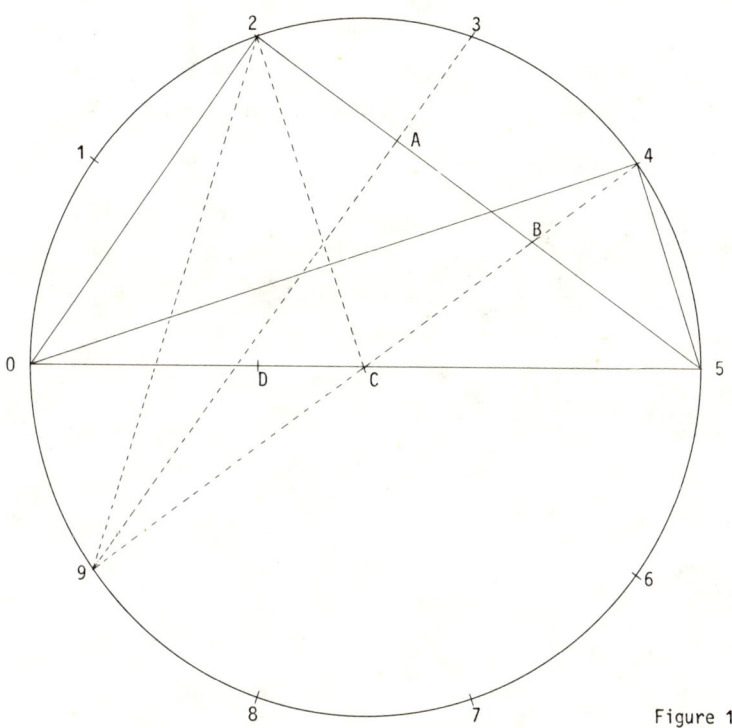

Figure 1

Note that A bisects 2B, 9A3 being height and median in B92; and note that 2B = 2C = radius, and that B5 = 45 = side of the decagon. Thus 25, the one that interests us, is the protagonist of XIII 9:

> "If the side of the hexagon and that of the decagon inscribed in the same circle be added together, the whole straight line has been cut in extreme and mean ratio, and its greater segment is the side of the hexagon".

For short: The point B divides 25 in the "Golden Section". Of which section we were told already in XIII 3 that "if a line segment be cut in extreme and mean ratio, the square on the sum of the lesser and half of the greater segment is five times the square on the half of the greater segment".

*In casu*:

$$\square\ \underline{5A} = 5\ \square\ \underline{A2} \tag{3˝}$$

Naming  a  any segment that is congruent with  $\underline{5A}$,
   r  any segment that is congruent with  $\underline{C2}$  (the radius),
and q  any segment that is congruent with  $\underline{A2}$  (a quarter of the diameter = half of the radius),

we write

$$\square\ a = 5\ \square\ q \tag{3}$$

Thus  $\underline{25}$  is a sum:  $\quad \underline{25} = \underline{5A} + \underline{A2} = a + q \tag{4}$

and  $\underline{45}$  is a difference:  $\underline{45} = \underline{5A} - \underline{BA} = a - q \tag{5}$

Writing a.q for "the rectangle that is contained by the segments a and q" (cf. II def 1, p. 19), we deduce with II 4 (p. 18):
Since  $\underline{25} = a + q$, $\qquad \square\ \underline{25} = \square\ a + \square\ q + 2\ a.q$ .
Substituting  $5\ \square\ q$  for  $\square\ a$, and  r  for  2q:

$$\square\ \underline{25} = 6\ \square\ q + a.r \tag{6}$$

Seeing that the diameter is 4q, we can rewrite (1)

$$\square\ \underline{02} + 6\ \square\ q + a.r = 16\ \square\ q$$

and subtracting  $6\ \square\ q$  we get

$$\square\ \underline{02} + a.r = 10\ \square\ q \tag{7}$$

To wit:  The square on the side of the pentagon is less than ten squares on the half of the radius by a rectangle which is contained by the radius and the side of the square that is five times the square on the half of the radius.

About the side of the decagon we deduce from (5) with II 7, *à la grecque* without using any minus signs:

Since  $\underline{45} + q = a$,    □ $\underline{45} + 2\,a.q =$ □ $a +$ □ $q$.

Substituting  5 □ q  for  □ a,  and  r  for  2q:

$$\square\ \underline{45} + a.r = 6\ \square\ q \tag{8}$$

Adding  □ $\underline{04}$, we get    □ $\underline{04}$ + □ $\underline{45}$ + a.r = 6 □ q + □ $\underline{04}$.

Substituting from (2), the  □ $\underline{05}$  being  16 □ q,  we write

$$16\ \square\ q + a.r = 6\ \square\ q + \square\ \underline{04},$$

and subtracting  6 □ q,  we end up with

$$10\ \square\ q + a.r = \square\ \underline{04} \tag{9}$$

To wit:  The square on the diagonal of the pentagon is greater than ten
squares on the half of the radius  by a rectangle which is contained by the radius  and  the side of the square that is five
times the square on the half of the radius.


0.2    The writer of Elements could perfectly well have pronounced and proved these statements about the squares on the side and diameter of the pentagon, which give exactly the information that we asked for in our initial paragraph; all the ingredients are present in the II'nd and in the XIII'th book, and our phrasing can bear comparison with II 13 and 14. Some theorems, as e.g. XIII 9, are very *apropos* under the assumption that such reasonings were done; and one theorem seems to be thrown in as a windfall, XIII 10: From equations (7) and (8) it appears that the difference between  □ $\underline{02}$  and  □ $\underline{45}$  is  4 □ q;  that is

The square on the side of a regular pentagon inscribed in a circle equals the sum of the squares on the side of the hexagon and on that of the decagon inscribed in the same circle.

That theorem was a starting point for Ptolemy when he calculated the table of chords in Almagest I 10. In the Elements it is proved elegantly by similar triangles and the general theory of proportion, avoiding numbers; the author, apparently, was wary with them. There has since long been consensus, as far as we know, that the theorem was hard to come by; but if some Greek reasoned like we did above, he could hardly miss it.

What strikes us in XIII 11, then, is the absence of numerical constants, which to us are vital parts of the theorem, and the obscurity of the language: What is a *rhētē* diameter? under what conditions is a line segment *álogos*? and the "Lesser" line segment is lesser than what? We shall elucidate, if not answer, those questions while presenting (in chapter 2) the lesson of the X'th book of the Elements. We believe that the numerical **constants** were deliberately suppressed by the author, seeing that they play an important part in his proof of the theorem:

Let the chord 28 intersect the diameter 05 in the point D; then OD is the projection on the diameter of the side of the pentagon 02. From a lemma about the rightangled triangle (X 32/33) we know that

the square on 02 equals the rectangle contained by the projection and the diameter: □ 02 = OD·O5 .

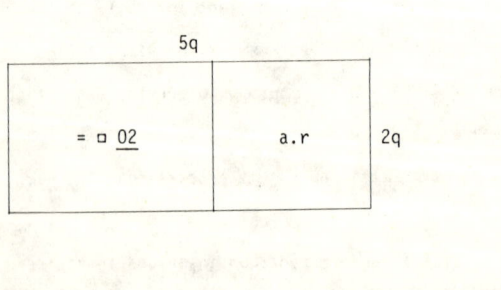
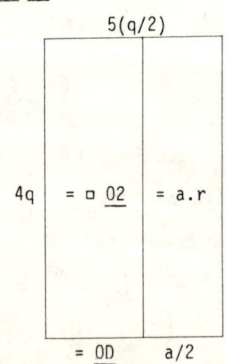

Figure 2

## Coloured Quadrangles

The square □ $O2$ and the diameter $O5$ being known, the projection $OD$ is easily determined. Figure 2 illustrates equation (7) in two ways:

10 □ q    can be represented by a rectangle contained by   2q   and   5q;
a.r       can be represented by a rectangle contained by   2q   and   a;
if a.r be subtracted from 10 □ q, the remainder represents □ $O2$.

If the same quadrangles be applied to the diameter, namely 4q, the widths will be 5(q/2) and a/2 **respectively**, and $OD$ is seen to be

$$5(q/2) - a/2.$$

Since  □ a = 5 □ q,        □ (a/2) = 5 □ (q/2).

And q/2 being a quarter of the radius, we can state that:

> $OD$ is less than five quarters of the radius by a line segment which is the side of the square that is five times the square on a quarter of the radius.

It is evident from XIII 11 that the author knew this much about $OD$, and that the proof depends on it. We do not *know* if Euclid or anybody else ever reasoned like this, but it is sufficiently manifest by now that it is no matter of difficult mathematics to determine $OD$ in terms of the radius.

And we are sufficiently motivated by now to have a look at the X'th book and find out about *rhēté* and The Lesser Line.

# Chapter 1

## Tools and Prerequisites

### 1.0 The Text.

The text of the Elements is well enough established by E. S. Stamatis *post* J. L. Heiberg (as he puts it) in the Teubner series. We suppose our reader to have access to that edition, or to some translation, e. g. the one by Thomas Heath. It will be easy, then, to check our statements and supply whatever we leave out.

### 1.1 Conventions and Notations.

The objects of our theory are

1) line segments, denoted by $\quad\quad\quad\quad a, b, c, \ldots, x, y.$

2) quadrangles, namely either squares, denoted by $\square a$, $\square b$, etc. or rectangles; the rectangle "contained by" the line segments $a$ and $b$ is denoted by $\quad\quad\quad\quad a.b.$

3) geometrical magnitudes, irrespectively of dimensions; they occur only in the general theory of commensurability (§ 1.4), and are denoted by $\quad\quad\quad\quad A, B, C,$ etc.

4) positive integers, denoted by $\quad\quad\quad\quad a, b, c,$ etc.

That $A$ is commensurable with $B$, is written $\quad\quad A$ *com* $B.$
That $A$ is incommensurable with $B$, is written $\quad\quad A$ *inc* $B.$
That the ratio of $A$ to $B$ is the same as that of $C$ to $D$, is written $\quad\quad\quad\quad (A : B) = (C : D).$

Theorems which are substantially identical with those of the Elements, are numbered as e.g. $\quad\quad\quad\quad$ II 4, VII 13, X 12, etc.

Theorems which are special cases of more general theorems of the Elements, or which are restated in our own idiom, are numbered as e.g.

T VI 1,   T VI 17,   etc.

At odd places in the X'th book *lemmata* (or: auxiliary theorems) prop up; they are sandwiched between theorems and numbered as e.g. X 13/14, etc.

Theorems and definitions of our own are numbered 201, 202, 203, etc.

Now and again we bracket our comments and notes in << ... >> , to make the theory stand out more clearly. Any inconsistencies in that respect should be borne patiently.


1.2   The premises.   Our Thesis of Simplicity.

The main theme of the X'th book commences with X 17. Before understanding that one, there are some things to be learnt: obviously, the theorems X 1-16, though we shall not deal explicitly with all of them; X 1 and 2, which are by far the best pieces of mathematics in the whole book, have very little to do with the rest of it. And there are some repetitions, because Euclid was in the habit of proving contrapositive theorems explicitly (e.g. X 7 and 8, 16).

Inasmuch as the X'th book, as we know it, is part of the whole corpus of Elements, we might simply take what propositions we need from the other books of the Elements. But it is a haunting impression (which haunted Oskar Becker already in the 1930's) that the Story of the Greater and the Lesser Line, viz. the X'th book, did not need Eudoxus' theory of proportion in order to be coherently told. Such an impression cannot be proved, perhaps not even made plausible; but we entertain the thesis

that it is possible to tell that story within a theory of ratio that is
confined to commensurable magnitudes, supplied by very few (less than 5)
"intuitively true" statements about the ratio of line segments and of
quadrangles, whether they be commensurable or no. We shall leave it as
an open question whether our thesis has any historical consequences.


1.3  The Second Book of the Elements.

The reasonings (which we shall render no less by figures than by words
and symbols) proceed along the lines taught in the second book; which,
by the way, had better begin at I 33 with the theory of parallelograms.
The second book may be interpreted in several ways; we propose the following:

If $a$ and $b$ be arbitrary line segments, $a > b$,
there is a uniquely determined couple of line segments, $s$ and $d$,
such that $a = s + d$ and $b = s - d$;
namely $s = (a + b)/2$ and $d = (a - b)/2$.

Now $a.b$, $\square a$, $\square b$, and their sums and differences can be expressed
in terms of $s$ and $d$, which is proved in II 4-10:

If $a = s + d$ and $b = s - d$, then

II 4        $\square a = \square s + \square d + 2\, s.d$ .

II 5 & 6    $a.b = \square s - \square d$ .

II 7        $\square b = \square s + \square d - 2\, s.d$ .

II 8        $\square a - \square b = 4\, s.d$ .

II 9 & 10   $\square a + \square b = 2(\square s + \square d)$ .


If book II were in the focus of this paper, we should discuss theorems

II 1, 2, and 3 as indicators of an arithmetical theory "underlying" the geometry of the book, and comment on the duplicity of II 5 & 6 and 9 & 10, as well as on the special difficulties concerning subtraction. But as it is not, we confine ourselves to the listing of the theorems.

Problems important to our subject are solved in II 11 and 14:

II 11   divides a given line segment in the Golden Section.
II 14   constructs a square equal to a given rectangle.

The book opens with a definition:

II def 1   Any rectangle is said to *be contained by* the two line segments that contain the right angle.

This is a curious terminology (which raised some discussion in the scholia), since a rectangle, in order to be contained, must have four sides. But its idea is to associate a rectangle with a couple of line segments, namely its unequal sides, and vice versa. Thoughts of "product" of line segments are tempting, but should be kept aside, as this is no theory of measuring, and the output: a rectangle, is a thing quite different from the line segments.

From book I we shall need the "Theorem of Pythagoras":

I 47 & 48   If in triangle ABC $\angle$ C is a right angle, then $\square\, a + \square\, b = \square\, c$.   And vice versa.

Proposition I 43 we quote in a special form concerning rectangles only, and including its converse, which is not in the Elements:

T I 43   In any rectangle the *complements* of the rectangles about the diagonal are equal.

T I 43 v.v.    Equal rectangles are complements in one
               and the same rectangle.

"Complements" translates the Greek *paraplērōmata*. In figure 3 the rectangles $R_1$ and $R_2$ are complements.

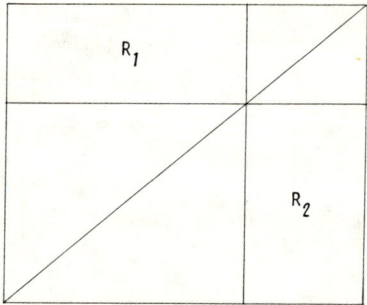

Figure 3

Together, those theorems constitute the essential criterium for equality of quadrangles. The vital construction of a rectangle with given height, equal to a given quadrangle, is established with the rectangular case of the "parabolical application of area":

T I 44    To a given line segment to apply (*parabalein*)
          a rectangle equal to a given quadrangle.

If $R_1$ be given, and the height of $R_2$, the width of $R_2$ can be found by "completing" the whole rectangle of figure 3.

## 1.4 Theory of Ratio and Commensurability.

Book X opens with the words "commensurable magnitudes", *sýmmetra megéthe*, but only the first definition and theorems 1-8, 11-13, and 15-16 are about geometrical magnitudes *in general*; the rest deal merely with line segments and rectangular quadrangles, as do definitions 2-4 and theorems 9, 10, and 14.

The very notion of commensurability rests on the undefined notion of measuring, *metrein*, which can be explained by the following statements:

"A measures B" is synonymous with "B is a multiple of A".

The first definition, X def 1, lays down that

"those magnitudes are said to be commensurable which are measured by the same measure, and those incommensurable which cannot have any common measure".

We shall employ the following notation of two magnitudes, A and B, of the same kind:

A *com* B   or   A *inc* B .

"Is commensurable with", *com*, is a relation in sets of magnitudes of the same kind. That it is not trivial, is established in X 1-2, which ensure the existence of incommensurable magnitudes, provided that there exist magnitudes, the alternative subtraction (*anthyphairesis*) of which does not come to an end. Such magnitudes are, e.g., the diagonal and the side of the regular pentagon.

However, the proofs of book X seldom have recourse to def 1, but to two theorems depending on it, namely

X 5   Commensurable magnitudes have to one another a ratio which a number has to a number.

And the converse:

X 6  If two magnitudes have to one another a ratio which a number has to a number, the magnitudes will be commensurable.

The transmitted proof of X 5 suffers from various defects. To our view, such a statement cannot be proved, but must be claimed: firstly, one has to establish a definition of proportionality of commensurable magnitudes, which can easily be done by imitating VII def 21, concerning proportional *numbers*. But, secondly, one must either postulate X 5 or define numbers as (a subset of) commensurable magnitudes (which may well be considered the same escape). We propose

Definition 200  Two magnitudes A and B have the same ratio as two numbers $a$ and $b$, if A and B have a common measure M which measures A $a$ times and B $b$ times.

Now X 5 follows immediately; let us put X 5 and 6 in symbols:

X 5 & 6  A *com* B ⟺ There exist natural numbers $a$ and $b$, such that (A : B) = ($a$ : $b$).

From X 5 and 6 follows X 11, by recourse to V 11 (see below):

X 11  If (A : B) = (C : D), then [A *com* B ⟺ C *com* D].

The theorem is problematic in a pre-Eudoxean context (cf. § 1.2), since we have not defined proportionality of magnitudes that are *not* commensurable; but we shall use it of such magnitudes only in connection with

T VI 1  Rectangles under the same height are to each other as their bases.

In figure 4 the rectangles h.a and h.b are to each other as their bases: $(h.a : h.b) = (a : b)$.

If h.a *com* h.b, then a *com* b; and if h.a *inc* h.b, then a *inc* b.

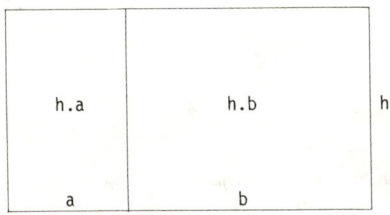

Figure 4

Proposition VI 1 is crucial to the theory of proportion in the set of plane polygons. We demand that it be granted us, without proof, in the rectangular form (which may well be the one that Aristotle alludes to in Topica 158 b 29 ff.)

Also without proof we accept some "intuitive" propositions, which were undoubtedly believed to hold for incommensurable as well as for commensurable magnitudes, even before Eudoxus:

T V 7   Equal quadrangles have the same ratio to one and the same quadrangle. And one quadrangle has the same ratio to equal quadrangles.

T V 9   Quadrangles which have the same ratio to one and the same quadrangle, are equal.

V 11    Ratios which are equal to one and the same ratio, are equal.

With those four theorems we can prove the following series from the V'th and VI'th book; we postpone our proofs and comments till § 3.4 in order not to delay our exposition:

VI 16   If four line segments be proportional, the rectangle contained by the extremes is equal to the rectangle contained by the means.               And vice versa.

In symbols: $(a : b) = (c : d) \iff a \cdot d = b \cdot c$.

A particular case of VI 16 is

VI 17     $(a : b) = (b : c) \iff a \cdot c = \square b$.

From VI 16 follows the *enallax*-theorem for line segments:

T V 16    $(a : b) = (x : y) \iff (a : x) = (b : y)$.

Wherewith the *di'isou*-theorem is easily proved:

T V 22    $\left. \begin{array}{l} (a : b) = (x : y) \\ (b : c) = (y : z) \end{array} \right\} \Rightarrow (a : c) = (x : z)$.

On this basis we can prove the "square cases" of VI 22 and VI 19:

T VI 22   $(a : b) = (c : d) \iff (\square a : \square b) = (\square c : \square d)$.

T VI 19   $(a : b) = (b : c) \iff (\square a : \square b) = (a : c)$.

- - -

The relation *com* is an *equivalence* relation. That it is reflexive and symmetrical, follows from its definition. That it is transitive, is stated in

X 12             $A \text{ com } C \wedge C \text{ com } B \Rightarrow A \text{ com } B$.

This is easily proved when realizing that different measures of one and the same magnitude are commensurable. By a *reductio in absurdum* we get

X 13             $A \text{ com } B \wedge A \text{ inc } C \Rightarrow B \text{ inc } C$,

which is equivalent with X 12. Lastly is proved an elementary theorem about addition and subtraction of commensurable magnitudes:

X 15    $A \text{ com } B \Rightarrow (A+B) \text{ com } A \wedge (A+B) \text{ com } B$.

$A \text{ com } B \Leftarrow (A+B) \text{ com } A \vee (A+B) \text{ com } B$.

X 16    $A \text{ inc } B \Rightarrow (A+B) \text{ inc } A \wedge (A+B) \text{ inc } B$.

$A \text{ inc } B \Leftarrow (A+B) \text{ inc } A \vee (A+B) \text{ inc } B$.

Among the theorems about magnitudes occurs a statement of line segments and their squares, in X 9: "The squares on line segments commensurable in length have to one another the ratio which a square number has to a square number. And squares which have to one another the ratio which a square number has to a square number, will also have their sides commensurable in length".      In our symbols it looks as follows:

X 9     $a \text{ com } b \Leftrightarrow$ There exist square numbers $a^2$ and $b^2$, such that $(\square a : \square b) = (a^2 : b^2)$.

The theorem is vital for the investigations in the X'th book, of which a greater part concerns squares that are commensurable while their sides are not. In the form transmitted to us the proof has raised much discussion; on p. 71 we present a simple proof of our own, based mainly on T V 9. For the moment we take its truth for granted and proceed to notice that as a consequence of X 5, 6, and 9 one and only one of the following statements holds for any couple of line segments a and b:

1) $\square a \text{ com } \square b \wedge a \text{ com } b$  (a is commensurable with b in length);
2) $\square a \text{ com } \square b \wedge a \text{ inc } b$  (a is commensurable with b in square only);
3) $\square a \text{ inc } \square b \wedge a \text{ inc } b$  (a is incommensurable with b in length and in square).

"In square" translates the Greek dative *dynámei*; about that word we have written in *Museum Tusculanum* 40-43 (Copenhagen 1980), pp. 120-131.  A précis of that paper can be read in appendix A, pp. 72-76.

Chapter 2

Coloured Quadrangles

2.0 <u>Why Colours? An initiation into not translating</u>.
In this chapter we pretend to set forth the essentials of the X'th book of Euclid's Elements, a theory of certain "irrational" line segments and quadrangles. One difficulty is to grasp the meanings and implications of the Greek adjectives *rhētós* "utterable", *mésos* "middle", and *álogos* "irrational". Another, having grasped them, is not to project into them what *we* know about surds and irrationals, and to refrain from arguing as if they were predicated of real numbers and not of lines and quadrangles.

In order not to convey to them any meaning before having experienced their range in the theory of the book, and seeing that those adjectives denote qualities of plane figures and line segments, we allow ourselves the linguistic trick of translating them in terms of colours, irrespective of the meaning we suppose they had to the Greeks. Thus, whenever the text has a form of *rhētós*, we write "red" (quite arbitrarly, from the very sound of it); *mesos* will be "amber" (and easy to remember if one thinks of the "middle" traffic light); and *álogos* "obscure".

The trick amounts to no more than *not* translating the adjectives, but borrowing them, like som many other Greek words; only it makes easier reading to use some current English words which are easy to remember. And since the mathematical properties of a *rhēté* line segment or a *rhētón* quadrangle are completely independent of the meaning of *rhētós* (as they depend only on the definitions of book X), we might for that matter translate the word by any English adjective so long as we use it according to those definitions. We hope to gain two advantages by the trick: To compel our reader to think of the definitions whenever the words are used. And of nothing else, that is: to suppress any prejudices about their meaning.

The output of the lesson will be a logic game with coloured quadrangles and line segments, leading to an unimportant classification of some line segments. Apart from the fun one can have (and some Greek surely had) from manipulating such objects, we shall not hesitate to maintain that the game, bar very few details, has no mathematical importance; the X'th book of the Elements may well be called a *cul-de-sac* in mathematics, even though it did inspire Kepler to his *Harmonice Mundi*. All the same, it is a fascinating, nay haunting, piece of literature, and its composition reminds one of epic or dramatic poetry, with its long sequences of uniform statements apt to be learnt by heart and transmitted orally. We have a few words to say about that in § 3.2.

As a petty contribution to Peace on Earth we shall refrain from polemizing against some current interpretations that we find - misleading. But we will be hard to convince that the X'th book of Euclid's Elements is not fairly interpreted by our coloured quadrangles.


2.1    Red Quadrangles and Red Line Segments.

Definition 201      There is a line segment  r,  which is set out
T X deff 3 & 4      beforehand. The segment  r  is *red*,  and its
                    square  □ r  is also *red*.

| | |
|---|---|
| Definition 202 | A quadrangle is *red* if and only if |
| T X def 4 | it is commensurable with □ r. |
| | |
| Definition 203 | A line segment is *red* if and only if |
| T X def 3 | it is the side of a red square. |

<< Those definitions are mere renderings of X deff 3 and 4. Please note that the line segment r, ἡ προτεθεῖσα εὐθεῖα, is no *unit* segment, as is often maintained. No measuring takes place in the X'th book, only *commensuratio*: every statement about a red line segment remains valid (even) if the segment is replaced by one that is commensurable with it. Thus r is the segment of comparison, or rather: the square □ r is the quadrangle of comparison.

"Being a red quadrangle" is synonymous with "being commensurable with the first square □ r"; which (X 5 & 6) means that "any red quadrangle has to the first square a ratio that an integer has to an integer". No fixed number is assigned to □ r; in fact, it plays the rôle of a phantom: it takes no part in the propositions, it does not appear in the figures, but serves merely to determine which quadrangles (and line segments) are red.

We might translate *rhētá* quadrangles by "rational" quadrangles; but we cannot translate *rhētaí* line segments by "rational" line segments, as the ratio of such ones will more often than not be that of a square root of an integer to a square root of a non-square integer. Correctly, then, we might name them "sides of rational squares", - and that is exactly what we want to do; only, to make easier reading, we write "red line segment" whenever we ought to write "side of a rational square". And to preserve the Greek uniformity of terms we also write "red" quadrangles for "rational" quadrangles. >>

Our next theorem collects some obvious consequences of 202 and 203, which are not explicit in the Elements, but are used *passim*:

Theorem 204      1) Red quadrangles are commensurable.

2) A quadrangle that is commensurable with a red quadrangle, is red.   A quadrangle that is incommensurable with a red quadrangle, is not red.

3) A sum of red quadrangles is red.

4) Red line segments are commensurable if and only if their squares have to each other the ratio which a square number has to a square number.

5) Incommensurable red line segments are said to be commensurable *in square only*.

6) A line segment that is commensurable with a red segment, is red.

Proofs:   1) and 2) follow from 202 and X 12 & 13.   3) follows from 204.1 and X 15.   4) and 5) are consequences of 203 and X 9.   6) follows from 203, 202, X 9 and 12.                                                                    □

Theorem 205                A rectangle which is contained by red line
T X 19                     segments  commensurable in length, is red.

Proof:      Let a rectangle  h.w  and the square  □ h  be drawn as in figure 5. From  T VI 1  we know that  (□ h : h.w) = (h : w).
Now, by hypothesis,  h *com* w,  whence  (X 11)  □ h *com* h.w.
And (203)  □ h  is red;  therefore  (204.2)  h.w  is red.            □

Figure 5

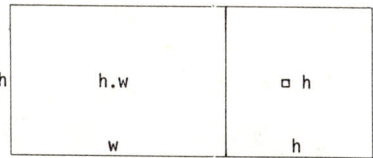

From now on, we shall time and again recall T I 44 (p. 20) with the phrase: To apply some quadrangle as a rectangle to some height ..., meaning: to construct a rectangle with that height, equal to the quadrangle in question.

Theorem 206  
T X 20

If a red quadrangle be applied as a rectangle to a red height, its width is also red and commensurable in length with its height.

Proof: Let h.w be a red rectangle with red height h, and let □ h be drawn as in figure 5.
Now both □ h and h.w are red and therefore (204.1) commensurable.
Since (T VI 1)    (□ h : h.w) = (h : w),
we deduce by X 11 that  h $com$ w;  whence (204.6)  w is red.    □

Theorem 207  
T X 21

A rectangle which is contained by red line segments commensurable in square only, is not red.

The Proof runs as 205 until
Now, by hypothesis, h $inc$ w, whence (X 11) □ h $inc$ h.w.
And (203) □ h is red; therefore (204.2) h.w is not red.    □

## 2.2    Amber Quadrangles and Amber Line Segments.

Definition 208  
T X 21

A quadrangle is *amber*, if and only if it is *equal to a rectangle* contained by red line segments commensurable in square only.

<< Note that "equal" is used in the sense of I 35; of course, it does not mean "congruent with". Our reasons for "translating" *mésos* by "amber" were expounded in § 2.0. *Mésos* indicates that the quadrangle in

## Coloured Quadrangles 31

question is a mean proportional between two red quadrangles which do not have a ratio to each other as two square numbers. The ratio of an amber quadrangle to a red one is that of a square root of a non-square integer to an integer; and the ratio of an amber line segment to a red one is that of a fourth root of a non-square integer to a square root of an integer. But you may forget that if you keep to the definitions. The Elements introduce the concept *méson* quadrangle rather confusedly: the word does not occur until the corollary of X 23, although the non-redness of a *méson* square is proved before that of its side in X 21. >>

Definition 209  A line segment is *amber*, if and only if
T X 21          it is the side of an amber square.

Theorem 210  If an amber quadrangle be applied as a rectangle
T X 22       to a red height, its width is also red but commensurable in square only with its height.

Proof:   Let  h.w  be an amber rectangle with red height  h.
Then  (208)  there is some rectangle  c.d = h.w,  with  c  and  d  red and commensurable in square only.
From  h.w = c.d  follows  (T VI 16)  that         (h : d) = (c : w),
whence (T VI 22, a drastic one!)         (□ h : □ d) = (□ c : □ w).
Both  □ h  and  □ d  are red,  and therefore commensurable  (204.1); thus (X 11)  □ c *com* □ w;  and  □ c  being red,  □ w  is also red; and so  w  is red, but commensurable in square only with  h, since  (205)  h.w,  by hypothesis, is not red but amber.         □

## Theorem 211

T X 23

A quadrangle which is commensurable with an amber quadrangle, is amber; and a line segment which is commensurable with an amber line segment, is amber.

Proof with figure 6:   Let □ a be (equal to) an amber quadrangle, and let some quadrangle (equal to) □ b be commensurable with □ a. Apply □ a and □ b to some red height c: c.d = □ a and c.e = □ b. Then (210) d is red and commensurable with c in square only. By hypothesis □ a *com* □ b, and consequently c.d *com* c.e . Since, by T VI 1, (c.d : c.e) = (d : e), we have (X 11) d *com* e. Therefore e is red and commensurable with c in square only (X 13). Thus c.e = □ b is amber.                                                               □

If a be an amber line segment, then (209) □ a is amber; if b *com* a then □ b *com* □ a. From 211 first part, □ b and so b is amber.   □

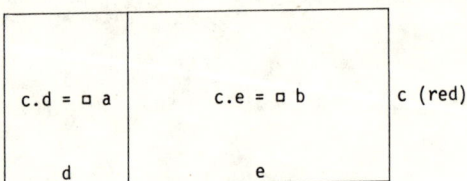

Figure 6

If a and b be amber line segments and (figure 6) □ a = c.d and □ b = c.e, with c, d, and e red, then (1) if d *inc* e, the amber squares will be incommensurable, and so *a fortiori* a and b. But (2) if d *com* e, the amber squares will be commensurable. And (2a) if (d : e) = ($d^2$ : $e^2$), with $d^2$ and $e^2$ square numbers, then a and b will be commensurable in length (X 9); (2b) otherwise, they will be commensurable in square only.    □

Thus we have observed that pairs of amber quadrangles may be incommensurable. As a matter of fact, most of them are.

Theorem 212      A rectangle which is contained by amber line
T X 24           segments commensurable in length, is amber.

The proof runs exactly as 205 if you read "amber" for "red", (209) for (203) and (211) for (204.2).                             □

Much later in the Elements, between X 53 and 54, a *lemma* is proved which will come in useful at this point of our treatise; (we venture a thesis on the position of *lemmata* in § 3.2, p. 66):
Applying T VI 1 twice on figure 7, first "horizontally", then "vertically", we have           (□ a : a.b) = (a : b)
and                               (a : b) = (a.b : □ b);
thus the rectangle a.b is the mean proportional to □ a and □ b:

Lemma 213        Any rectangle is the mean proportio-
X 53/54          nal between the squares on its sides.

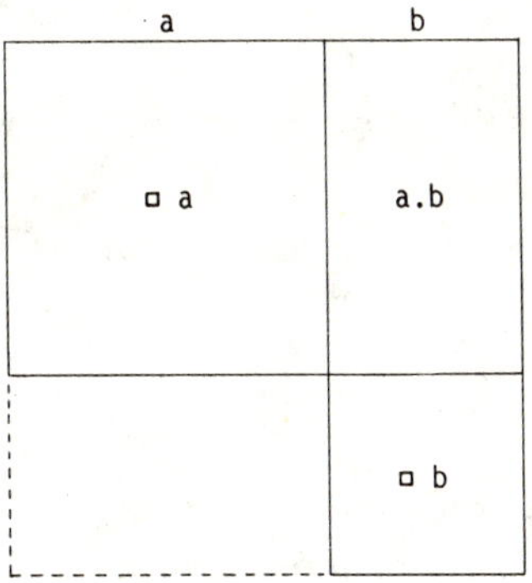

Figure 7

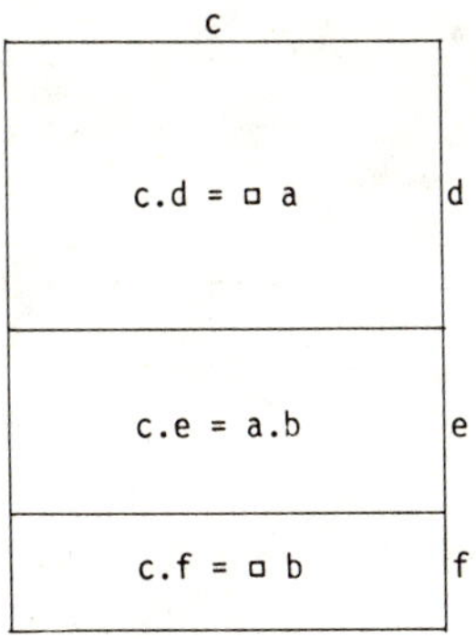

Figure 8

Theorem 214   A rectangle which is contained by amber line
T X 25        segments commensurable in square only,
              is either red or amber.

Proof with figures 7 and 8:   Let a and b be amber line segments, commensurable in square only.
Apply □ a, a.b, and □ b to some red line segment c:

   c.d = □ a,   c.e = a.b,   and   c.f = □ b.

By theorem 210, d and f are red; and they are commensurable, because
(d : f) = (c.d : c.f) = (□ a : □ b),   and by hypothesis □ a $com$ □ b.
Since a.b is the mean proportional to □ a and □ b (213),
the segment e must be the mean proportional to d and f; therefore
(VI 17) □ e = d.f, and (205) □ e is red, whence (203) e is red.
If e $com$ c,  then       c.e, that is: a.b is red; but
if e $inc$ c,  then                    a.b is amber.               □

The converse of 212 and 214 are not in the Elements; we state them here, leaving their proofs to our reader as an exercise:

Theorem 215   If a red rectangle has an amber height, its width is
              also amber and commensurable in square only with its
              height.
              If an amber rectangle has an amber height, its width is
              also amber and either commensurable in lenght with its
              height or commensurable in square only with its height.

From 214 and 215 and figures 7 and 8 it is obvious how to find amber line segments commensurable in square only which contain a red rectangle (X 27), and such which contain an amber one (X 28); in X 55 & 56 they manifest themselves by what appears as another route, though it is not:

A first amber line segment  a  is found from an arbitrary couple of red
line segments  c  and  d,  commensurable in square only:  viz. the mean
proportional  a  to  c  and  d;  then  □ a = c.d,  an amber quadrangle.
Let  e  be any third red line segment, commensurable in square only with
d;  and use  T I 44  to find  b  such that  a.b = c.e.
[Then, by  T VI 1,    (a : b) = (□ a : a.b) = (c.d : c.e) = (d : e),
so the trick can be described as finding  b  from the ratio
(a : b) = (d : e) ].
Now, if  e *com* c,  then  c.e,  that is:  a.b  is red   (205) *.
And  if  e *inc* c,  then                 a.b  is amber  (208).         □

* The construction in X 27 takes  e = c,  not merely  e *com* c;  thus its
generality is very much limited, although of course it produces what it
promises to do. (For checking, you may need a concordance:  A = d,
B = c,  Γ = a,  Δ = b.)


Of line segments, commensurable in square only, we have until now come
across three types of pairs:

       1) Red segments which contain an amber rectangle   (208);

       2) Amber segments which contain a red rectangle    (214);

       3) Amber segments which contain an amber rectangle  (214).

All of them are to play significant rôles in our next paragraph, to
which we premise

Definition 216   Quadrangles that are not red or amber, are *obscure*.
T X 3 & 4        A line segment is obscure if and only if it is the
                 side of an obscure square.

"Obscure" translates *álogos* "irrational". We keep to our colours.

## 2.3 Obscure Quadrangles. Sums and Differences of Line Segments.

We proceed to investigate sums and differences of line segments which are commensurable in square only; in accordance with 203 and 209 we study the squares on such sums and differences:

Theorem 217  
T X 36

The sum of two red line segments commensurable in square only, is obscure. Let it be called "Sum-of-two-reds".

Proof: Let $x$ and $y$ be red line segments commensurable in square only, and put $x + y = q$.

Then (II 4, with figure 10, left) $\square q = \square x + \square y + 2 x.y$ .

The sum $\square x + \square y$ is red (204.3), and the rectangle $x.y$ is amber (208), whence (211) $2 x.y$ is also amber and incommensurable with the red sum $\square x + \square y$ (indirectly from 204.2).

Therefore $\square q$ is not red (X 16).

[ Nor is it amber; for the difference between two amber quadrangles ($\square q$ and $2 x.y$) is not red (*infra*, 219). ]

Thus (216) $\square q$ is obscure and $q$ is obscure. $\square$

Theorem 218  
T X 73

The difference between two red line segments commensurable in square only, is obscure. Let it be called "Apotome".

Proof: Let $x$ and $y$ be red line segments commensurable in square only, and put $x - y = q$.

Then (II 7, with figure 11, top) $\square q + 2 x.y = \square x + \square y$ .

The proof goes on as 217 until ...

[ Nor is it amber; for the sum of two amber quadrangles is not red (*infra*, 219). ] Thus $\square q$ is obscure and $q$ is obscure. $\square$

*Coloured Quadrangles* 37

<< The arguments in brackets [] in 217 and 218 are proved otherwise in
the Elements X 72/73 and 111/112; systematically they belong here.
"Sum-of-two-reds" translates the Greek *ek dýo onomáton* "from two names".
If it must be latinized, it should be *binominalis*, not *binomialis*, as
tradition has it. Did someone lose the nasal stroke over *-ia-* ?
We borrow the Greek *apotomé* "which is cut off", *i.e.* "difference". >>

Theorem 219          The sum of two amber quadrangles is not red.
T X 26              The difference between two amber quadrangles
                     is not red.

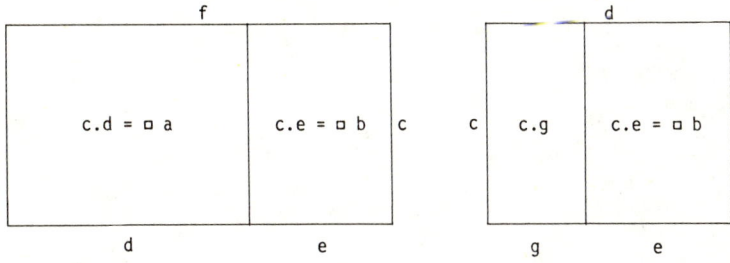

Figure 9

Proof with figure 9: Let the amber squares □ a and □ b, a > b, be
applied to some red line segment c:        □ a = c.d and □ b = c.e .
Then d and e are red and commensurable with c in square only (*).
1) Suppose the sum □ a + □ b = c.f to be red; then c and f are red
and commensurable in length, whence (*) d and f are red and commensurable in square only. But their difference e is red, though it was
proved in 218 that it is *not* red; which is absurd.
Therefore □ a + □ b is not red.                                   □

2) Suppose the difference □ a - □ b = c.g to be red; then c and g are red and commensurable in length, whence (*) e and g are red and commensurable in square only. But their sum d is red, though it was proved in 217 that it is *not* red; which is absurd.
Therefore □ a - □ b is not red.    □

(*) Those conclusions are ensured by 210, 206, and X 13.
<< X 26 deals only with the *difference* of amber quadrangles (*mésa*); we have reconstructed the other part, for the sake of symmetry.
Arithmeticians will realize that if r (of 201) has a rational length, at least one of the segments x and y must have a square root for its length; and □ a and □ b must have square roots for their areas. They may maintain that we have proved in 217 and 218 that
$x \pm \sqrt{y}$ and $\sqrt{x} \pm \sqrt{y}$ are neither rational nor square roots (red), nor fourth roots (amber);   and in 219 that $\sqrt{a} \pm \sqrt{b}$ is not rational.
In an arithmetical context these would be (parts of) one theorem; but in the X'th book they are and must be separate propositions about different kinds of magnitudes. Compare p. 67, please.    >>


## 2.4 Partition of Equal Quadrangles and Elliptical Application of Areas.

The squares on "Sums-of-two-reds" and on "Apotomes" having been proved obscure, we shall apply such squares to a red height, in order to determine the colour of the width of the rectangles (compare 206 and 210).
To that end we prove, once and for all, two theorems about partitions of equal quadrangles, *in casu*: a square □ q and a rectangle h.w = □ q; they are distilled from the repetitive theorems X 54-65 and 91-102, about which more will be said in § 2.6:

Theorem 220  If $\square q = h.w$, and if $q = x + y$, with $x > y$,
then $w$ is a sum of such segments, $w = b + c$, $b > c$,
that $h.b = \square x + \square y$ and $h.c = 2 x.y$.

Conversely: If $h.w = \square q$, and if $w = b + c$, with $b > c$,
then $q$ is a sum of such segments, $q = x + y$, $x > y$,
that $\square x + \square y = h.b$ and $2 x.y = h.c$.


Theorem 221  If $\square q = h.w$, and if $q = x - y$, then $w$ is a
difference between such segments, $w = b - c$,
that $h.b = \square x + \square y$ and $h.c = 2 x.y$.

Conversely: If $h.w = \square q$, and if $w = b - c$, then $q$ is a
difference between such segments, $q = x - y$,
that $\square x + \square y = h.b$ and $2 x.y = h.c$.


Proof of 220 first part, with figure 10:  If $q = x + y$, then (II 4) $\square q$ can be divided into four parts: $\square x + \square y + 2 x.y$.
Since $\square q = h.w$, all four parts must fit into $h.w$ somehow. So if we construct $h.u = \square x$, $h.v = \square y$, and put $u + v = b$, so that $h.b = \square x + \square y$, then the remainder, which we name $h.c$, must equal $2 x.y$. And $b > c$, because $(\square x + \square y) > 2 x.y$ (X 59/60 or II 7). $\square$

Proof of 221 first part, with figure 11:  If $q = x - y$, then (II 7) $\square q + 2 x.y = \square x + \square y$; thus the sum $(\square x + \square y)$ is divisible into three parts: $\square q + 2 x.y$.  If, as in 220, we construct $h.u = \square x$, $h.v = \square y$, and put $u + v = b$, we have $h.b = \square x + \square y$. If from that rectangle we subtract $h.w$ ($= \square q$), then the remainder, which we name $h.c$, must equal $2 x.y$.  And $w = b - c$. $\square$

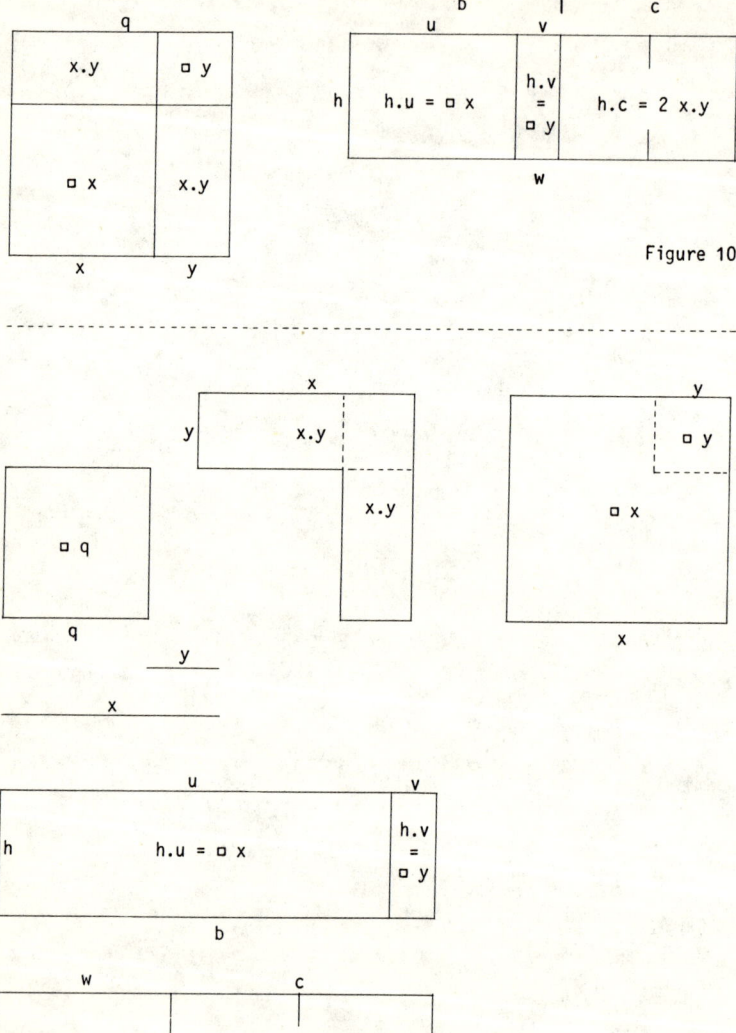

Figure 10

Figure 11

## Coloured Quadrangles

<< The obstacle of the converse parts is to get h.b split into such parts h.u and h.v that h.u = □ x and h.v = □ y, while h.c = 2 x.y.  Analysis reveals that, as x.y is the mean proportional to □ x and □ y (213), so half of h.c is the mean proportional to h.u and h.v.

But then (T VI 1) c/2 is the mean proportional to u and v, whence (VI 17)     u.v = □ (c/2) = 1/4 of □ c.

Thus b is split into unequal segments u and v which contain a rectangle equal to a given square. This situation would be recognized by any Greek geometer as the so-called *elliptical application of areas*. The phrasing can be learnt from X 17:

> "If there be two unequal line segments [b > c], and if to the greater a rectangle [u.v] be applied equal to one fourth of the square on the less [1/4 of □ c, that is: □ (c/2)] and falling short (*elleîpon*) by a square [□ v] ... "

This is the rectangular case of the general elliptical application of areas, which is solved in VI 28, where a condition of possibility is stated. The method of solution for the rectangular case is taken for granted in X 17, from which we elicit the following doings (figure 12):

If a rectangle u.v equal to 1/4 of □ c is applied to b
and falling short by the square □ v,
then b is cut into unequal segments u and v.
Let it also be cut into equal segments, either of them named b/2, and name the segment between the two points of section d/2.

Then II 5 teaches us that          □ (b/2) = u.v + □ (d/2),
and since by hypothesis             u.v = □ (c/2),
we have                             □ (b/2) = □ (c/2) + □ (d/2).

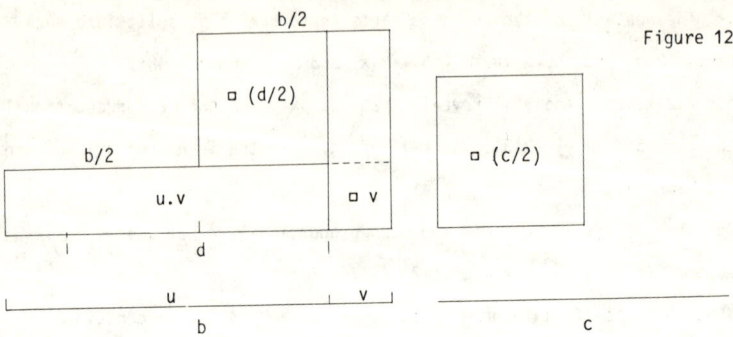

Figure 12

Multiplying by 4 we get □ b = □ c + □ d.  (*)
As b and c are given, d is given by way of a rightangled triangle.
Whence the segments u and v are given as

u = b/2 + d/2   and   v = b/2 - d/2.

In the idiom of Book X the equation (*) says that "b masters more than
(*dýnatai meîzon*) c by the square on d"; as it turns out to play an
important role, we shall name d "the square difference" between b
and c, and denote it   d = √(□ b - □ c),   without forgetting that
d is a line segment, not a number in any sense of the word.   >>


Proof of 220 converse part, with figure 10:    Let b be cut into
b = u + v (with u > v), such that u.v = 1/4 of □ c. (Elliptical application of area).    Construct □ x = h.u   and   □ y = h.v .
Now x.y must equal h.c/2, as they are mean proportionals to respectively equal quadrangles.    So  2 x.y = h.c .
Put x + y = q´ (+). Then (II 4) □ q´ = □ x + □ y + 2 x.y
                                  = h.b   + h.c = h.w = □ q .
Thus q = q´ = x + y.  And  x > y, because u > v, since b > c.    □

The proof of 221 converse part, with figure 11, is identical with 22o, if you read in line 6:

Put $x - y = q^{\check{}}$ (†). Then (II 7) $\square\, q^{\check{}} = \square\, x + \square\, y - 2\, x.y$ . etc.

<<(†) X 54-59 and 91-96 *construct* a square equal to h.w by constructing $\square\, x$ and $\square\, y$ about one and the same diagonal. We represent that by the trick with $q^{\check{}}$. >>

## 2.5 Six Classes of Sums-of-two-reds and Apotomes.

In what follows we shall be constantly referring to the figures 10 and 11 and state our theorems in terms of those figures, speaking of $w_+$ and $q_+$ in figure 10 and of $w_-$ and $q_-$ in figure 11. We begin with

Theorem 222     If u *com* v, then b *com* d, where d is
T X 17 & 18     "The square difference"  $\sqrt{(\square\, b - \square\, c)}$ .
                And conversely.

The proof is obvious from X 15, since (figure 12)
$u = b/2 + d/2$,  $v = b/2 - d/2$, and  $b = u + v$,  $d = u - v$.  $\square$

The purport of the theorem is that conditions of commensurability between u and v, and therefore between $\square\, x$ and $\square\, y$, can be stated in terms of b and c, and conversely.

Theorem 223     If the square on a sum-of-two-reds be applied
T X 60 & 97     as a rectangle to a red height, then the width
                of the rectangle is a sum-of-two-reds.
                And if the square on an apotome be applied as
                a rectangle to a red height, then the width of
                the rectangle is an apotome.

In terms of figure 10 and 11:

If x and y are red line segments commensurable in square only, then b and c will be red line segments commensurable in square only.

Proof: Since □ x and □ y are red, their sum is also red (204.3); that is: h.b is red, and (206) b is red and h $com$ b.
And since x and y are red and commensurable in square only, x.y is amber (208), and so (211) 2 x.y is amber; therefore c is red (210) and commensurable with h in square only. Whence b $inc$ c. Thus $w_+$ = b + c is a sum-of-two-reds, and $w_-$ = b - c is an apotome.

□

<< Does the converse hold? If b and c be red, commensurable in square only, and if they be applied to a red height h, will h.u and h.v be red, as they must in order to make □ x and □ y red?
Answer: If and only if u and v are commensurable with h and consequently with each other, and (X 15) with b. Whence, by virtue of 222, b $com$ d. And, of course, b $com$ h.
But even then: will h.c be amber, so as to make x.y amber and x and y commensurable in square only?
Answer: It will, because c is commensurable with h in square only. Thus h, b, and c must be very special sets of reds, if the converse of 223 is to be valid:   >>

<u>Theorem 224</u>   If h, b, and c are red line segments, and b commensurable with c in square only, then the following six classes exist:

*Coloured Quadrangles* 45

1. class:   h *com* b  ∧  h *inc* c  ∧  b *com* d = √(□ b - □ c).

2. class:       *inc*           *com*           *com*

3. class:       *inc*           *inc*           *com*

4. class:       *com*           *inc*           *inc*

5. class:       *inc*           *com*           *inc*

6. class:       *inc*           *inc*           *inc*

<< Those are the logical possibilities for combining the relevant conditions of commensurability (cf. the "Second and Third Definitions", X 47/48 and 84/85). That they do exist, can be proved by construction and appeal to X 9. Squares in the ratio of given numbers can be constructed as in X 6/7, which is based on T VI 19:

>   Construct line segments  a  and  c  in the given ratio  (a : b);
>   find the mean proportional  b  to  a  and  c.

From  (a : b) = (b : c)  follows  (□ a : □ b) = (a : c) = (a : b). The numbers mentioned are positive integers; they are square numbers if *and only if* they have the exponent ². >>

Problem 224.1   The first class is not empty:

T 48 & 85       Take four numbers  $h^2$  and  $b^2 = c + d^2$.                    (†)

Let  h  be a red line segment, and construct  b  and  c such that   (□ h : □ b : □ c) = ($h^2$ : $b^2$ : c).

Then  (X 9)  h *com* b  and  h *inc* c  and  b *inc* c.
And  (□ b : □ d) = ($b^2$ : ($b^2$-c)) = ($b^2$ : $d^2$). Whence  b *com* d.       □

An example:    $h^2 = 4$,  $b^2 = 9$,  $c = 5$,  and  $d^2 = 4$.

<< (†) That there are square numbers  $b^2$  and  $d^2$  which do not differ by a square number, can be proved by establishing conditions under which they *do* (cf. lemma X 28/29 about the "Pythagorean Triple"), and observe

## Problem 224.2   The second class is not empty:

T X 49 & 86   Take four numbers $h^2$ and $b^2 = c + d^2$.

Let $h$ be a red line segment, and construct $b$ and $c$ such that   $(\square h : \square b : \square c) = (h^2 : b^2 c : c^2)$.

Then $h$ *inc* $b$ and $h$ *com* $c$ and $b$ *inc* $c$.   And $b$ *com* $d$, because $(\square b : \square d) = (b^2 c : (b^2 c - c^2)) = (b^2 c : d^2 c) = (b^2 : d^2)$.   □

An example:   $h^2 = 4$, $b^2 = 4$, $c = 3$, and $d^2 = 1$.

<<   Lines 4 and 6 look somewhat more sophisticated than the text of X 49 and 86; but then, Euclid has cloaked the intricacy of his transactions, which amount to our construction.   >>

## Problem 224.3   The third class is not empty:

T X 50 & 87   Take four numbers $h$ and $b^2 = c + d^2$.

Let $h$ be a red line segment, and construct $b$ and $c$ such that   $(\square h : \square b : \square c) = (h : b^2 : c)$.

Then $h$ *inc* $b$ and $h$ *inc* $c$ and $b$ *inc* $c$.

And $b$ *com* $d$, because $(\square b : \square d) = (b^2 : (b^2 - c)) = (b^2 : d^2)$.   □

An example:   $h = 2$, $b^2 = 9$, $c = 5$, and $d^2 = 4$.

## Problem 224.4   The fourth class is not empty:

T X 51 & 88   Take four numbers $h^2$ and $b^2 = c + d$.

Let $h$ be a red line segment, and construct $b$ and $c$ such that   $(\square h : \square b : \square c) = (h^2 : b^2 : c)$.

Then $h$ *com* $b$ and $h$ *inc* $c$ and $b$ *inc* $c$.

And $b$ *inc* $d$, because $(\square b : \square d) = (b^2 : (b^2 - c)) = (b^2 : d)$.   □

An example:   $h^2 = 4$, $b^2 = 9$, $c = 6$, and $d = 3$.

Problem 224.5    The fifth class is not empty:

T X 52 & 89    Take four numbers $h^2$ and $b = c^2 + d$.

Let h be a red line segment, and construct b and c such that    ($\square$ h : $\square$ b : $\square$ c) = ($h^2$ : $b$ : $c^2$).

Then h *inc* b and h *com* c and b *inc* c.

And b *inc* d, because ($\square$ b : $\square$ d) = ($b$ : ($b - c^2$)) = ($b$ : $d$).    $\square$

An example:    $h^2 = 4$, $b = 3$, $c^2 = 1$, and $d = 2$.

Problem 224.6    The sixth class is not empty:

T X 53 & 90    Take four numbers h and $b = c + d$.

Let h be a red line segment, and construct b and c such that    ($\square$ h : $\square$ b : $\square$ c) = ($h$ : $b$ : $c$).

Then h *inc* b and h *inc* c and b *inc* c.

And b *inc* d, because ($\square$ b : $\square$ d) = ($b$ : ($b - c$)) = ($b$ : $d$).    $\square$

An example:    $h = 2$, $b = 5$, $c = 3$, and $d = 2$.


The decomposition of the set of sums-of-two-reds and of apotomes in 224 is a *Class Decomposition*: That every sum-of-two-reds and every apotome pertain to at least one of the six classes, is obvious. That it pertains to no more than one, is a consequence of the following theorem:

Theorem 225    The partial segments of a sum-of-two-reds

T X 42 & 79    and of an apotome are uniquely determined.

Proof (1): Suppose that $q = x + y = u + v$ are different decompositions of one and the same sum-of-two-reds, with $x > y$ and $x > u \geqslant v$. Then (II 4)    $\square$ q = $\square$ x + $\square$ y + 2 x.y = $\square$ u + $\square$ v + 2 u.v .
Now, the difference between the two sums of squares must be the same as the difference between the two double rectangles, the former being red

while the latter is not red (219); which is absurd.

Therefore there can be no difference, and $x + y$ and $u + v$ are not different decompositions of q.

(2) Suppose that $q = x - y = u - v$ are different decompositions of one and the same apotome.

Then (II 7) $\square q + 2 x.y = \square x + \square y$, and $\square q + 2 u.v = \square u + \square v$.

Now, the difference ... as in (1), *mutatis mutandis*. □

<< If a segment $q (= x - y)$ is given to be an apotome, the Elements have very special terms for x and y, namely for y: *prosharmózousa* "The One that Fits", and for x: "The Whole". Theorem 225 says:

> To an apotome there is only one red segment that fits
> and is commensurable in square only with the Whole.

A marvellous theorem if you come to think of it. >>

We are able to amend theorem 223:

<u>Theorem 223 bis</u>  If the square on a sum-of-two-reds be applied as a
T X 60 & 97      rectangle to a red height, then the width of the
                 rectangle is a *first* sum-of-two-reds.
                 And if the square on an apotome be applied as a
                 rectangle to a red height, then the width of the
                 rectangle is a *first* apotome.

All that we need to add is the observation (in figures 10 and 11) that $u \; com \; v$, whence $b \; com \; d = \sqrt{(\square b - \square c)}$ . □

This amendment takes us immediately to the questions:
If the width w shifts through the six classes of sums-of-two-reds or apotomes, what will happen to the square side $q = x \pm y$ ?

## 2.6 The Greater, the Lesser, and their Family.

Theorems 226-231 will answer those questions; their proofs are very uniform, if not tedious, and the reasonings are more easily grasped by a diagram (p. 50). The rectangle of the first sum-of-two-reds (resp. the first apotome) equals a square the side of which is a sum-of-two-reds (resp. an apotome), whereas the others do not. We believe *that* insight to be the aim of the whole theory.

The other square sides (which may be said to be of a further order than the sums-of-two-reds and apotomes) have hardly any rôles to play, but retaliate by bearing noble titles: they are named, if possible, according to their partial segments, but else according to the decomposition of their squares; except two of them, the *Greater* and the *Lesser*, which cannot but attract our attention.

We are aware of some verbosity of the theorems, though we have reduced the arguments to a minimum. They are meant to verbalize the diagram on p. 50, and will turn quite simple if one thinks of the figures 10 and 11 while memorizing them. The first is the converse of 223 bis:

| | |
|---|---|
| Theorem 226 | If $w_+$ is a first sum-of-two-reds, then ▫ $q_+$ is a sum of two red squares and an amber rectangle, and thus $q_+$ is a sum-of-two-reds. |
| T X 54 & 91 | |
| 36 & 73 | |
| | And if $w_-$ is a first apotome, then ▫ $q_-$ is a difference between a sum of two red squares and an amber rectangle, and thus $q_-$ is an apotome. |

The thing is obvious from line 1 of our diagram: $x$ and $y$ are red and commensurable in square only, because their rectangle is amber (210). Thus they are well-known segments from 217 and 218. ▫

| | A | B | C | D | E | F | G | H | GoTo line No. |
|---|---|---|---|---|---|---|---|---|---|
| | w | u & v | | | □ x | □ y | □ x + □ y | 2 x.y | |
| | b ± c | b & d | h & b | h & c | h.u | h.v | h.b | h.c | |
| 1.class | *com* | *com* | *inc* | | red | red | red | amber | 1 |
| 2. - | *com* | *inc* | *com* | | amber$_1$ | amber$_1$ | amber$_1$ | red | 2 |
| 3. - | *com* | *inc* | *inc* | | amber$_1$ | amber$_1$ | amber$_1$ | amber$_2$ | 3 |
| 4. - | *inc* | *com* | *inc* | | | | red | amber | 4 |
| 5. - | *inc* | *inc* | *com* | | | | amber | red | 5 |
| 6. - | *inc* | *inc* | *inc* | | | | amber$_1$ | amber$_2$ | 6 |

---

| line No. | K<br>name of $q_+$ = x + y | L<br>name of $q_-$ = x - y | Theorems |
|---|---|---|---|
| 1 | sum-of-two-reds | apotome | 226<br>X 54, 91, 36, 73 |
| 2 | first sum-of-two-ambers | first apotome-of-two-ambers | 227<br>X 55, 92, 37, 74 |
| 3 | second sum-of-two ambers | second apotome-of-two-ambers | 228<br>X 56, 93, 38, 75 |
| 4 | Greater, *Meízōn* | Lesser, *Elássōn* | 229<br>X 57, 94, 39, 76 |
| 5 | Mistress of a red and amber quadrangle | Maker of an amber whole together with a red | 230<br>X 58, 95, 40, 77 |
| 6 | Mistress of two amber quadrangles | Maker of an amber whole together with an amber | 231<br>X 59, 96, 41, 78 |

## Coloured Quadrangles

The diagram p. 50 should be read in the terms of figures 10 and 11:

$\square\, q = h.w$ .

$q_+ = x + y \iff \square\, q_+ = \square\, x + \square\, y + 2\, x.y$

$\qquad\qquad\qquad\qquad\quad = h.u + h.v + h.c$

$\qquad\qquad\qquad\qquad\quad = \quad h.b \quad + h.c \;=\; h.w_+$ .

$q_- = x - y \iff \square\, q_- + 2\, x.y = \square\, x + \square\, y$

$\qquad\qquad\qquad\; h.w_- + h.c \;=\; h.u + h.v \;=\; h.b$ .

The line segments h, b, and c are red, b and c commensurable in square only. Remember that u and v are "solutions" to the elliptical application of area $\quad u + v = b \quad \& \quad u.v = \square\,(c/2)$,

and that $\qquad\qquad\quad u\; com\; v \iff b\; com\; d = \sqrt{(\square\, b - \square\, c)}$ .

Amber quadrangles of one and the same line are commensurable if and only if they have the same index.

How to use the diagram:

If w be given as one of the classes of column A, the corresponding items of columns B, C, and D are given by 224; whence the colours of columns G and H are determined by 205 and 208, that is: the colours of $h.b = \square\, x + \square\, y$, and of $h.c = 2\, x.y$.

It so happens that the sum of squares and the double rectangle have different colours in lines 1, 2, 4, and 5. Therefore (X 16 and 219) the squares $\square\, q_+ = \square\, x + \square\, y + 2\, x.y$ and $\square\, q_- = \square\, x + \square\, y - 2\, x.y$ are neither red nor amber.

In lines 3 and 6, both $\square\, x + \square\, y$ and $2\, x.y$ are amber, but incommensurable, because b *inc* c and so h.b *inc* h.c (T VI 1, X 11). Their sum $\square\, q_+$ and difference $\square\, q_-$ is neither red (206) nor amber (210). Therefore all the line segments of columns K and L are obscure (which is proved rather prolixly in X 36-41 and 73-78).

The colours of x and y are determined, if u *com* v, that is: if
b *com* d = √(□ b - □ c), as the rectangles h.u and h.v are then commensurable with h.b and must have its colour (204.2, 211).

If u *inc* v, the colour of x and y is left undetermined, being outside the scope of the apparatus in the X'th book.

We shall speak of the sets of obscure line segments in the lower part of our diagram as K 1- K 6 and L 1- L 6, for short; their names are introduced in 227 - 231. □

<u>Theorem 227</u>  If $w_+$ is a second sum-of-two-reds, then □ $q_+$ is a sum
T X 55 & 92  of two commensurable amber squares and a red rectangle;
  37 & 74  thus $q_+$ is a sum of two amber line segments commensurable in square only which contain a red rectangle. It is obscure and should be called "*First sum-of-two-ambers*".

If $w_-$ is a second apotome, then □ $q_-$ is a difference between a sum of two commensurable amber squares and a red rectangle, and thus $q_-$ is a difference between two amber line segments commensurable in square only which contain a red rectangle. It is obscure and should be called "*First apotome-of-two-ambers*".

Proof: From line 2 of our diagram we learn that x and y must be amber; they are commensurable in square only, because their rectangle is red (215). □

<u>Theorem 228</u>  If $w_+$ is a third sum-of-two-reds, then □ $q_+$ is a sum of
T X 56 & 93  two commensurable amber squares and an amber rectangle
  38 & 75  incommensurable with them; thus $q_+$ is a sum of two amber

line segments commensurable in square only which contain
an amber rectangle. It is obscure and should be called
"*Second sum-of-two-ambers*".

If $w_-$ is a third apotome, then □ $q_-$ is a difference between a sum of two commensurable amber squares and an amber rectangle incommensurable with them; thus $q_-$ is a difference between two amber line segments commensurable in square only which contain an amber rectangle. It is obscure and should be called "*Second apotome-of-two-ambers*".

Proof from line 3: As in 226 and 227 □ x *com* □ y (because u *com* v), but □ x *inc* x.y (because u *inc* c/2); thus x and y are commensurable in square only. They are amber, because their squares are commensurable with h.b, which is amber. □

Theorem 229   If $w_+$ is a fourth sum-of-two-reds, then □ $q_+$ is a sum
T X 57 & 94    of two incommensurable squares the sum of which is red,
   39 & 76     and an amber rectangle; thus $q_+$ is a sum of two line
               segments incommensurable in square which contain an amber
               rectangle while the sum of their squares is red.
               It is obscure and should be called "*Greater*".

If $w_-$ is a fourth apotome, then □ $q_-$ is a difference between a sum of two incommensurable squares the sum of which is red, and an amber rectangle; thus $q_-$ is a difference between two line segments incommensurable in square which contain an amber rectangle while the sum of their squares is red. It is obscure and should be called "*Lesser*".

It is all evident from line 4 of our diagram, except the naming. □

Theorem 230   If $w_+$ is a fifth sum-of-two-reds, then $\square\, q_+$ is a sum of
T X 58 & 94    of two incommensurable squares the sum of which is amber,
     40 & 77   and a red rectangle; thus $q_+$ is a sum of two line seg-
               ments incommensurable in square which contain a red rect-
               angle while the sum of their squares is amber. It is obs-
               cure and should be called "*Mistress of a red and amber*
               *quadrangle*".

If $w_-$ is a fifth apotome, then $\square\, q_-$ is a difference between an amber sum of two incommensurable squares and a red rectangle; thus $q_-$ is a difference between two line segments incommensurable in square which contain a red rectangle while the sum of their squares is amber. It is obscure and should be called "*Maker of an amber whole together with a red*".

Even the naming is evident from line 5 of our diagram. "Mistress" translates *dynaméne* "who masters", which may also be rendered "square side". $\square\, q_-$ makes an amber whole ($\square\, x + \square\, y$) together with a red ($2\, x.y$). $\square$

Theorem 231   If $w_+$ is a sixth sum-of-two-reds, then $\square\, q_+$ is a sum of
T X 59 & 95    two incommensurable squares the sum of which is amber,
     41 & 78   and an amber rectangle; thus $q_+$ is a sum of two line seg-
               ments incommensurable in square which contain an amber
               rectangle while the sum of their squares is amber. It is
               obscure and should be called "*Mistress of two amber*
               *quadrangles*".

If $w_-$ is a sixth apotome, then $\square\, q_-$ is a difference between an amber sum of two incommensurable squares and an amber rectangle; thus $q_-$ is a difference between two line segments incommensurable in square which contain an amber rectangle while the sum of their squares is amber. It is obscure and should be called "*Maker of an amber whole together with an amber*".

Proof from line 6 of our diagram. The two amber quadrangles (□ x + □ y) and 2 x.y are incommensurable, because b *inc* c.  □

<< Because the sets K 2-6 and L 2-6 are *defined* in terms of the partitions of □ q and therefore of h.w, the converse theorems of 227-231 are trivial. We take 229 converse as a paradigm: >>

Theorem 229 conv.   Let q be a *Greater*.
T X 63              Then, *by definition*, □ q is a red sum of two in-
                    commensurable squares (□ x + □ y) and an amber
                    double rectangle (2 x.y).

When applied to a red height h, the square □ q equals h.w, which can be split into h.u = □ x, h.v = □ y, and h.c = 2 x.y.
Then h.u *inc* h.v, whence u *inc* v, and so b *inc* d = √(□ b − □ c).
And h.u + h.v = h.b is red, whence h *com* b,
while h.c is amber, whence       h *inc* c.
Thus w = b + c is a *fourth sum-of-two-reds*.   □

Let us recapitulate 226-231 with their converse theorems:

$$w_+ \text{ is an } i\text{'th sum-of-two-reds} \iff q_+ \in K\, i$$
$$w_- \text{ is an } i\text{'th apotome} \iff q_- \in L\, i$$

From 225 follows that if w be given, b and c are uniquely determined; whence (for fixed h) h.u, h.v, h.b, and h.c are uniquely determined, and so x and y, and q, are uniquely determined.

From 233 (which we proceed to prove) follows the converse: that if $q$ be given, $b$ and $c$, and $w$, are uniquely determined.
Thus we have established the counterpart of X 54-65 and 91-102:

<u>Theorem 232</u>    There is a one-to-one correspondence between the $i$'th sum-of-two-reds and K $i$, and between the $i$'th apotome and L $i$.

<u>Theorem 233</u>    The line segments of columns K and L on p. 50
T X 43-47          are uniquely split.
   80-84

Proof: Suppose that some $q$ of K $i$ has two decompositions: $q = x + y = x_1 + y_1$, with $x > x_1$.
Then  $\square q = \square x + \square y + 2 x.y = \square x_1 + \square y_1 + 2 x_1.y_1$ = h.w.
From lemma 234 (*infra*)    $\square x + \square y > \square x_1 + \square y_1$;
but then h.w has two different decompositions,

   h.w = h.b + h.c = h.$b_1$ + h.$c_1$ , with $b > b_1$.

Which is impossible (225). Therefore $q$ cannot have two decompositions. The same holds of any $q$ of L $i$.    $\square$

<u>Lemma 234</u>    $q = x + y = x_1 + y_1 \wedge x > y \wedge x > x_1 > y_1 \Rightarrow$
X 41/42       $\square x + \square y > \square x_1 + \square y_1$.

Proof: Put $x + y = x_1 + y_1 = 2 s$;   $x - y = 2 d$;   $x_1 - y_1 = 2 d_1$.
Then $x = s + d$ and $x_1 = s + d_1$.     From $x > x_1$ follows $d > d_1$.
Now (II 5, cf. § 1.3)    $x.y + \square d = \square s$
                    and $x_1.y_1 + \square d_1 = \square s$.
Whence $x.y < x_1.y_1$, and so $2 x.y < 2 x_1.y_1$.
Since $\square x + \square y + 2 x.y = \square x_1 + \square y_1 + 2 x_1.y_1$, we conclude
   $\square x + \square y > \square x_1 + \square y_1$.    $\square$

From the fact that the subdivision of 224 is a class-decomposition follows that the sets K 1-6 and L 1-6 are disjunct (X 70 & 111 postscripts). However, to complete the case, we must prove

Theorem 235                        An apotome is not a sum-of-two-reds.
T X 111

Proof: Suppose $q = x + y$ (a sum-of-two-reds) $= x_1 - y_1$ (an apotome). Apply $\square$ $q$ to a red height $h$; then $w = b + c$ is a first sum-of-two-reds, and $w = b_1 - c_1$ is a first apotome (figures 10 and 11).
Thus $b$ and $b_1$ are commensurable with $h$,
and $c$ and $c_1$ are incommensurable with $h$.
Put $b_1 - b = e$; then (X 15) $e$ $com$ $h$ and $e$ $inc$ $c$.
Now $e = c + c_1$, whence $c = e - c_1$, an apotome; but $c$ is also red, which is impossible. Therefore an apotome is not a sum-of-two-reds. $\square$


It is part of the structure that commensurable segments pertain to the same set. We have known that much about red and amber segments for some time, but it also holds that, e.g., a segment which is commensurable with a Greater, is itself a Greater. We refer to the text of the Elements for those truths, relying on Thomas Heath, who wrote in his commentary to X 66 and 103: "The proofs of (those) propositions ... are easy and require no elucidation". They presuppose

X 14         If $(b : c) = (b_1 : c_1)$, and if $b$ $com$ $d = \sqrt(\square\ b - \square\ c)$,
                then $b_1$ $com$ $d_1 = \sqrt(\square\ b_1 - \square\ c_1)$.             And vice versa.

Which is a trivial consequence of T VI 22 and a few well-established rules for handling proportions.

C U R T A I N

## 2.7 Postlude: Friendship between Sums and Apotomes.

The curtain falls, as it were, after the establishment of theorem 232, leaving the audience in some bewilderment as to the point of the play. We are prepared to face the possibility that there was no other point than to entertain us with good logic; we shall return to the question in the last chapter, after mentioning that piece of good mathematics which appears as X 112-114, concerning the connexion between sums-of-two-reds and apotomes, different though they be:

T X 112   If a red quadrangle be applied as a rectangle to a sum-of-two-reds, the width of the rectangle will be an apotome the terms of which are commensurable with the terms of the sum-of-two reds and in the same ratio; and the apotome is in the same class as the sum-of-two-reds.

T X 113   is identical with T X 112 if one reads "apotome" for "sum-of-two-reds" and vice versa.

T X 114   If a rectangle be contained by an apotome and a sum-of-two-reds, and if the terms of the sum be commensurable with and in the same ratio as the terms of the apotome, then the rectangle will be red.

Thomas Heath, in his commentary on X 112, wrote: "Like so many proofs in Archimedes and Apollonius, it leaves us completely in the dark as to how it was evolved. That the Greeks must have had some analytical method which suggested the steps of such proofs seems certain; but *what* it was must remain apparently an insoluble mystery".

We are not sure that they *must* have had some analytical method, - but

## Coloured Quadrangles

they surely had a way of tackling such problems without lapsing into mere trial and error; we must allow for a tendency among geometers to apply quadrangles to line segments whenever they felt like it, e.g.: If there be two unequal squares, to apply either of them to the sum or difference of their sides. We shall see it at work in the following analysis of T 112 and 113, where we try to follow closely the text's brilliant (but in no way far-fetched) juggling with proportions. For the sake of clarity we have printed them parallelly on p. 61; we suggest that you return to this point after reading that page. ...

In either case, $z$ is red, and therefore

in T 112   $y = (y+z) - z$  is an apotome,

and in T 113   $x - z + (x-z)$ is a sum-of-two-reds.

The terms are in the ratio $x:y$, that is $b:c$ (cf. equation (1). Why is $b.y = c.x$ ?). Whence follows *via* X 14 that the class of $y$ (resp. $x$) has the same number as the class of $b+c$ (resp. $b-c$). □

The top-hat trick is to find $z$ of equation (2), (STAMATIS p. 203, line 5 γεγονέτω); not the slightest hint is given as to how it was contrived, but then - we are not in the slightest doubt that it was done by applications of areas in the way we suggest.

T X 114 can be proved much easier than the text does it, with figure 15:

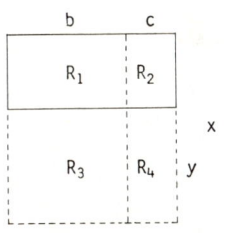

Figure 15

Let $b+c$ be a sum-of-two-reds, and let $x-y$ be an apotome, with
$(b : c) = (x : y)$ and $b$ *com* $x$ ∧ $c$ *com* $y$.
Now, $R_3 = R_2 + R_4$   (since $b.y = c.x$);
and $R_1 + R_3 = b.x$ is red;  substituting:
$R_1 + R_2 + R_4$ is red; and $R_4 = c.y$ is red;
therefore (X 15) $R_1 + R_2$ is red. □

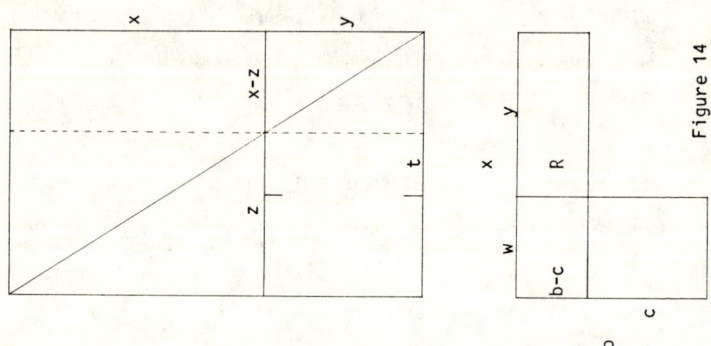

Figure 14

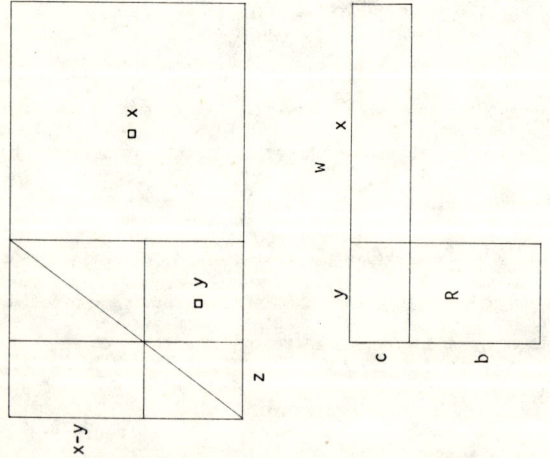

Figure 13

T X 112, with figure 13: Let a red quadrangle R be applied as a rectangle to b+c, a sum-of-two-reds, and let its width be y.   Then y is not red.
But it is part of a red segment $w = x+y$;   for if R be applied to c, its width w is red.
The idea suggests itself that $y$ *might be an apotome*.
Now $b.y = c.x$, whence $(x : y) = (b : c)$  (*1)
Thus x and y are commensurable in square (only), and so x is not red (because y is not red).
Unfortunately, $y = w-x$ is no apotome, so far.
But we can *make* y a difference between two segments commensurable in square only, by finding z such that
$$(y+z : z) = (x : y) \quad (*2)$$
Διελόντι (V def 15), (*2) is equivalent with $(y : z) = (x-y : y)$, whence $\square y = z.(x-y)$.
That is: if $\square y$ be applied to x-y, the width of the rectangle will be z.
From (*2) is evident that z and y+z are commensurable in square only; we want them to be red:
$$(*2) \iff (x+y+z : y+z) = (y+z : z) \quad \text{(by V 12)},$$
whence, according to T VI 19
$(x+y+z : z) = (\square (y+z) : \square z) = (\square x : \square y)$.
Therefore x+y+z *com* z, and so $x+y (= w)$ *com* z.

T X 113, with figure 14: Let a red quadrangle R be applied as a rectangle to b-c, an apotome, and let its width be x.   Then x is not red.
But part of it is a red segment $w = x-y$;   for if R be applied to b, its width w is red.   The idea suggests itself that x *might be a sum-of-two-reds*.
Now $b.y = c.x$, whence $(x : y) = (b : c)$  (**1)
Thus x and y are commensurable in square (only), and so y is not red (because x is not red). Unfortunately, $x = w+y$ is no sum-of-two-reds, so far.
But we can *make* x a sum of two segments commensurable in square only, by finding z such that
$$(z : x-z) = (x : y) \quad (**2)$$
Συνθέντι ἀνάπαλιν (V deff 13 & 14), (**2) is equivalent with $(z : x) = (x : x+y)$, whence $\square x = z.(x+y)$.
That is: if $\square x$ be applied to x+y, the width of the rectangle will be z.
From (**2) is evident that z and x-z are commensurable in square only; we want them to be red:
$(**2) \iff (x-z : y-(x-z)) = (z : x-z)$  (by V 19), which when we rename $(x-z : t) = (z : x-z)$, whence
$(z : t) = (\square z : \square (x-z)) = (\square x : \square y)$.
Therefore z *com* t, and so $z-t$ ($= x-y = w$) *com* z.

Chapter 3

By Way of Commentary

3.1     Genesis of the X'th Book: The Regular Pentagon?

It is a matter of dispute who was the author of the X'th book. We shall give our attention to another question: what kind of problem may have inspired the invention of the twice six sets of curious *álogoi* line segments (p. 50) ? We shall never *know*, but suppose one got interested in the regular decagon with its ten equal sides and no less than four different sizes of diagonals (figure 1), one of which is the diameter.

The side of the decagon, and the three kinds of diagonals, two of which happen to be the side and diagonal of the pentagon - can they be spoken of in terms of the diameter (which becomes thereby *rheté*, speakable *par excellence*, "red") ?

Let the squares on those diagonals be calculated, as in equations (6) to (9), pp. 12 and 13; from the very outset of the calculation (equation 4) it is evident that 25 is a sum of segments the squares on which can be spoken of in terms of the diameter: a quarter of the diameter plus the side of five squares on a quarter of the diameter.

Can that expression be simplified?

No! (X 36). But to answer that question we need a coherent theory of commensurability. Let it be invented. Thus we can take it for proved that the diagonal 25 is a sum of two square sides that cannot be reduced into one, but can only be spoken of separately:  it must have two names, be *ek dýo onomáton* (cf. p. 37). And that 45, the side of the decagon, is a difference, an *apótome* in the jargon of the day, between the same square sides.

A glance at equations (7) and (9) would suggest that the squares on the diagonal and side of the pentagon are of the same kinds of sum and difference as those just established; but if they be so, we must be able to split 10 □ q into two commensurable squares that will make up a complete square together with a.r, that is 2 a.q. We are landed in the analysis of theorem 220, p. 41:

If h.b = 10 □ q and h.c = 2 a.q, to split b into u + v with u.v = 1/4 of □ c, and determine if u *com* v.

(We are driven to prove X 17, 222.)

For h we may choose the diameter 05, being red; w = b ± c will then become the projection (on the diameter) of 04 and 02 respectively (lemma X 32/33), cf. figure 2 on p. 14.

Then b = 5 (q/2) and c = a/2

whence □ b = 25 □ (q/2) and □ c = 5 □ (q/2),

and so □ d = □ b - □ c = 20 □ (q/2). Thus (□ b : □ d) = (5 : 4), which happens not to be the ratio of a square number to a square number. Therefore b *inc* d, and so u *inc* v, and b is not divided into two parts u and v which are commensurable with h.

We have recognized the conditions for a line segment to be a sum or difference of two segments the squares on which are commensurable with the square on the diameter (223 bis and 226; cf. our considerations p. 44.) And we have found out that the squares on the side and diagonal of the pentagon, though being respectively a sum and difference of a red and an amber quadrangle, have sides of a "further order" than 25 and 45, and than their own projections on the diameter. Which observation will easily lead to the classification of 224 and entail the rest of Book X.

Connoisseurs will recognize the reasonings of XIII 11 in the above analysis. What *is* in fact proved in XIII 11 is this:

If an equilateral pentagon be inscribed in a circle which has a red diameter, the side of the pentagon is obscure of the kind that corresponds to a 4'th apotome (as described in 229). Line segments of that kind are called "Lesser" (because the Ancients named the side of the pentagon "The Lesser Line of the Pentagon" ?)

The history of XIII 11 may be the following:

1   The compositions of the (squares on the) side and diagonal of the equilateral pentagon were discovered. Those segments were called (for fun ?) the Lesser and the Greater (Line of the Pentagon).

2   They were proved to belong to distinct well-defined sets of *álogoi* line segments.

3   Those sets were named after them: The (set of the) Lesser and The (set of the) Greater.

4   The reasons for naming those sets were suppressed or forgotten. But the names persisted together with the definitions of the characteristics of the sets.

5   Thus it became necessary to *prove* that the side of the pentagon is the so-called Lesser (and that the diagonal is the so-called Greater; which theorem has not come through to us in the Elements).

We believe that some Greek mathematicians loved to give mathematical concepts odd names that were not self-evident to outsiders: πρῶτος ἀριθμός, δύναμις, λόγος, just to mention a few. We would not put it past them to name "Greater" any segment that has a splitting similar to that of the Greater Line of the Pentagon, to wit: the diagonal, and so on. That would be quite in line with the covering up of the trend of thought which makes the X'th a closed book:

## 3.2 Crux Mathematicorum? Or the By-Heart-Way of Learning?

"La difficulté du dixième livre ... est à plusieurs devenue une horreur, voir jusqu'à l'appeler *le croix des mathématiciens*, matière trop dure à digérer et en laquelle (ils) n'aperçoivent aucune utilité", wrote Simon Stevin in 1585 (quoted from B. L. van der Waerden *Science Awakening*, p. 172).

As indicated above (p. 58) we are in a strait to explicate to what end the investigation has brought us, though we have endeavoured to make our steps stand out clearly: beginning with a red square and some theorems about red quadrangles and red line segments, commensurable as well as incommensurable, *via* amber quadrangles and line segments to sums and differences of incommensurable red line segments. Which turned out to be classified into twice six classes and to be intimately connected with five "further" sets of irreducible sums and differences of line segments, some of which got odd names, whereas others had everyday names as the Greater and the Lesser. Lastly we saw that a sum is another thing than a difference of incommensurable reds, though they are connected by a red rectangle.

In the Elements the facts are presented somewhat obscurely and long-windedly in an upside down way: Not until X 36 do we meet the main object of research, when the *ek dýo onomáton* is defined and proved to be *álogos*. Then, in X 37-41, five other sums of peculiar line segments are defined from the blue, as it were, without any suggestion as to their significance, and proved to be *álogoi*; that they do exist is ensured with X 29-35, which seemed rather un-called for at first reading, and that they are uniquely split into their parts is proved in 42-47 (one truth of lasting interest). Then the six classes of *ek dýo onomáton* are defined and proved to exist, and in X 54-59 we learn that each of the six classes of *ek dýo onomátōn* leads to one and only one of the six *álogoi* of X 36-41. After that the author proceeds to prove the converses to X 54-59, and then *mirabile dictu*: the whole performance is given once again with apotomes, from X 73 onwards.

We have got the idea that the X'th book has some features in common with drama and with oral poetry: it is remarkably divided into sequences ("acts" as it were) of similar propositions and similar arguments, able to be learnt by heart even if dimly understood; and kept apart by "choral songs": *lemmata* or preliminary theorems, which are not always necessary nor in their proper systematic place, as if they were meant primarily to separate the "acts".

We cannot but applaud van der Waerden's judgement, that "the author succeeded admirably in hiding his line of thought". It may have been part of the teaching: if one knows (as we do) what it is all about, the order of the Elements is as easy to follow as any didactic and pedagogic presentation, perhaps even easier to learn by heart; and whoever is not initiated may be more easily kept outside by the opacity of the theme. A peculiar instance of Μ Η Δ Ε Ι Σ   Ε Ι Σ Ι Τ Ω .

## 3.3   An unauthorized arithmetical interpretation.

According to II def 1 there is some relation between a rectangle and the line segments that contain it; but that relation is defined nowhere in the Elements, and we know that it cannot be given without *measuring* and ressort to the field of *real* numbers, both alien to the Elements. Therefore the following interpretation is unauthorized from its very outset:

## Coloured Quadrangles

Let $\alpha(\square\, q)$ denote the *area* of the quadrangle $\square\, q$, and $|q|$ the *length* of the line segment q. (Thus $\alpha(\square\, q)$ and $|q|$ are *numbers*.) We define, unauthorizedly, area as a product of lengths:

A 1     $\alpha(h.w) = |h|\cdot|w|$, and particularly $\alpha(\square\, q) = |q|\cdot|q|$.

A 2     Whence, if $\alpha(\square\, q) = q$, then $|q| = \sqrt{q}$.

Let numbers $s\sqrt{h}$ and $t\sqrt{h}$, with $s$ and $t$ rational, be called *homonymous* square roots.

Now we are "entitled" to interpret the main statements of book X as follows (though we shall fatigue nobody with "translating" the proofs):

201-204     A quadrangle is red, *rhetón*, if and only if it has a rational area. A line segment is red, *rheté*, if and only if its length is a square root of a rational number.

X 19 & 20     The product of square roots of rational numbers is rational if and only if they are homonymous square roots.

208     A quadrangle is amber, *méson*, if and only if its area is a product of non-homonymous square roots of rational numbers which are not both rational *squares*.

Thus     the area of a *méson* quadrangle is a square root of a non-square rational number.
The side of a *méson* square is a *mése* line segment; its length is a fourth root of a non-square rational number.

Arithmetically, the areas of *mésa* quadrangles are included in the lengths of *rhetaí* line segments, being square roots of rationals: e.g. X 26 proves "the same thing" as X 73. In a geometrical theory this is obvious nonsense; there is strict maintenance of dimensions in the book.

About the sum and difference of square roots of rationals we learn in

X 36 & 73   If $x$ and $y$ are rational, not both squares, then $\sqrt{x} \pm \sqrt{y}$ is neither rational nor a square root of a rational.   (And we added: nor a fourth root of a rational.)

The theorem of unique splitting of irrationals is proved in

225   If $\sqrt{x} \pm \sqrt{y} = \sqrt{u} \pm \sqrt{v}$, then $x = u$ and $y = v$.

The main series of theorems, X 54-65 and 91-102, which we schematized on p. 50 and illustrated by figures 10 and 11, may be shortened into one about the square root of a sum or difference of square roots:

X 54 & 60
91 & 97
   Let $|h| = 1$ (some loss of generality!), $|b| = \sqrt{b}$ and $|c| = \sqrt{c}$, with $b$ and $c$ rational, but not both squares. Then $\sqrt{\sqrt{b} \pm \sqrt{c}} = \sqrt{x} \pm \sqrt{y}$, with $x$ and $y$ rational, if and only if $\sqrt{b-c}$ and $\sqrt{b}$ are rational.

The rest of the series study the logical alternatives, which have little arithmetical interest, though one can exert oneself (and one's mini-computer) in sorting out the six types of sums and differences from the set of couples of integers.

The "Theorems of Friendship" are well-known:

X 112 & 113           $a/(\sqrt{b} \pm \sqrt{c}) = k(\sqrt{b} \mp \sqrt{c})$.

Note that there is no question of *calculating* $k = a/(b - c)$.

X 114   The number $k \cdot (\sqrt{b} - \sqrt{c}) \cdot (\sqrt{b} + \sqrt{c})$ is rational if $b$, $c$, and $k$ are rational.

We leave out of this survey the general theory of commensurability (X 1-16), which though it may be considered preliminary to the study of quadrangles, has a much wider scope than most of the X'th book.

Authorized or not, an arithmetical version is, at best, unsatisfactory; the theme is likely to barter its fascinating geometric charm and clearness against some uninteresting absurdities if it be conceived, not to say: "proved" as arithmetic. A non-mathematical friend of ours once insisted that he still remembered the Theorem of Pythagoras from his schooldays. When asked to demonstrate it, he wrote on the blackboard:

$$a^2 + b^2 = c^2.$$

It took some time and several Socratic questions to make him draw a right-angled triangle and relate the equation to that! Some treatments of the X'th book remind us of that friend of ours.

We are tempted to adopt Plato's inscription on his door:

$$A\ G\ E\ O\ M\ E\ T\ R\ \bar{E}\ T\ O\ S \quad M\bar{E}DEIS\ EISIT\bar{O}$$

which is normally rendered "let no one enter who does not know geometry". But who can tell that he did not mean to say

$$\text{Arithmeticians Keep Out.}$$

3.4      Appendix A.   Some proofs.

On p. 24 we stated some theorems from the V'th and VI'th book of the Elements, without proofs. Before proceeding to prove them, we call attention to theorems V 16 and 22, the *enallax*- and *di'isou*-theorem, which we prove for line segments only, whereas in the Elements they are proved for magnitudes in general, - by far a more difficult task.    And VI 22 and 19 are proved here of squares only, not of any polygons.   We limit ourselves to the level that is necessary to ensure the theorems of the X'th book.

Proof of VI 16:
If  (a : b) = (c : d),  then  (T VI 1)   (a.d : b.d) = (b.c : b.d); whence  (T V 9)         a.d = b.c.
Conversely:   If  a.d = b.c,    then      (a.d : b.d) = (b.c : b.d), by virtue of  T V 7.   Whence  (T VI 1)     (a : b) = (c : d).         □

VI 17 follows immediately by putting  c = b  and renaming  d  as  c.   □

In  T V 16,  the left side proportion holds if and only if    a.y = b.x; but so does the right side proportion. Which settles it.           □

Proof of T V 22:
If  (a : b) = (x : y)  ∧  (b : c) = (y : z),  then  *enallax*
    (a : x) = (b : y)  ∧  (b : y) = (c : z);  whence   (V 11)
    (a : x) = (c : z),  and  *enallax*  (a : c) = (x : z).          □

Proof of T VI 22:
Let  (a : b) = (c : d).    Then   (T VI 16)   a.d = b.c.

Put $\square a = b.x$, whence (T VI 1) $(\square a : \square b) = (x : b)$.  (1)

Put $\square c = d.y$, whence $(\square c : \square d) = (y : d)$.  (2)

But $(a : d) = (\square a : a.d) = (b.x : b.c) = (x : c)$.  (3)

And $(b : c) = (b.c : \square c) = (a.d : d.y) = (a : y)$.  (4)

From (3) appears that (T VI 16) $a.c = d.x$ and from (4) that $a.c = b.y$.

Thus $d.x = b.y$, whence $(x : b)^{\bullet} = (y : d)$, which means, *via* (1) and (2) $(\square a : \square b) = (\square c : \square d)$.  □

Conversely: Let $(\square a : \square b) = (\square c : \square d)$.
Make $(a : b) = (c : z)$. Then $(\square a : \square b) = (\square c : \square z)$, as was proved above. But, by hypothesis, $(\square a : \square b) = (\square c : \square d)$. Whence $\square z = \square d$, and so $z = d$. Thus $(a : b) = (c : d)$.  □


Proof of T VI 19:
If $(a : b) = (b : c)$, then $a.c = \square b$ and $(\square a : \square b) = (\square b : \square c)$, whence $(\square a : \square b) = (a.c : \square c) = (a : c)$.  □


Of the vital theorem X 9 we give a proof depending on T VI 22. In the Elements, the proposition is proved with VI 22, but involving some defects, as pointed out by Thomas Heath in his commentary. How it was in fact proved by Theaitetos (who invented it, according to the scholiast) we dare not say. Our proof depends ultimately on V 9 or *de facto* V 8:

Proof of X 9:
If $a$ *com* $b$, then there is some line segment $e$ which measures both $a$ and $b$; let $e$ measure $a$ $\alpha$ times and $b$ $\beta$ times, with $\alpha$ and $\beta$ integers. Now counting will persuade us that

$\square$ a contains $a^2$ squares equal to $\square$ e   and

$\square$ b contains $b^2$ squares equal to $\square$ e.

Therefore, according to def 200,     ($\square$ a : $\square$ b) = $(a^2 : b^2)$   $\square$

Conversely (inspired by the method of X 6):

Let ($\square$ a : $\square$ b) = $(a^2 : b^2)$, with $a^2$ and $b^2$ square integers.

Let a be divided into $a$ equal segments, each to be named e.

Then $\square$ a contains $a^2$ squares equal to $\square$ e.

Let c be the line segment that is measured by e $b$ times;

then (a : c) = (a : b), and $\square$ c contains $b^2$ squares equal to $\square$ e.

Therefore (200)     ($\square$ a : $\square$ c) = $(a^2 : b^2)$.

But, by hypothesis, ($\square$ a : $\square$ b) = $(a^2 : b^2)$.

Therefore (V 9) $\square$ b = $\square$ c, and so b = c. And (a : b) = (a : b).   $\square$

## 3.5  Appendix B.  The word *dýnamis*.

Line segments a and b are said to be *dynámei sýmmetroi*, if $\square$ a is commensurable with $\square$ b. If, furthermore, a *inc* b, the segments a and b are *dynámei mónon sýmmetroi*, "commensurable in square only". For convenience we have adopted Thomas Heath's rendering, although we believe the word *dýnamis* to mean something else; consider the following analogy:

1) A line segment, εὐθεῖα γραμμή πεπερασμένη, has the vital property of having "extension in one dimension", μῆκος.

2) A limited plane surface, χωρίον ἐπίπεδον, has the vital property of having "extension in two dimensions", δύναμις.

3) A spatial magnitude, στερεόν μέγεθος, has the vital property of having "extension in three dimensions", ὄγκος.

If the *real* numbers are at hand, the "extensions" or "capacities" can always be subjected to measuring by a unit, so that such numbers are associated with them, known as the *length* of the segment, the *area* of the surface, and the *volume* of the spatial magnitude.

3.51   To understand why we do not immediately recognize the concept of *dynamis*, it will be instructive to note the three levels of the above analogy:

1)    a.    line segment.
     b.    extension in one dimension, *mēkos*.
     c.    number of length.

2)    a.    limited plane surface.
     b.    extension in two dimensions, *dýnamis*.
     c.    number of area.

3)    a.    spatial magnitude.
     b.    extension in three dimensions, *ó(n)gkos*.
     c.    number of volume.

Modern mathematics does not need level b, as it does not care which "capacity" is being measured, but employs functions of length, area, or volume to map the three kinds of magnitudes into the set of positive real numbers. Once the number is arrived at, we may forget which "medium" we are measuring. But to the Greeks, whose concept of number was limited to rational numbers, not every plane surface could be associated with a number, seeing that not every plane surface can be measured by a given unit. Never the less, every plane surface has an "extension", which one can speak of by calling it the *dynamis* of the surface.

3.52.   However, the word is conspicuously absent from most books of the Elements, occurring only in the X'th and the XIII'th book. In the doctrine of geometrical magnitudes, which is taught in books I-VI and

XI-XII, the concept is superfluous: level b is discarded together with c: If one cannot measure, it is unnecessary to speak of "something" that cannot be measured. The presence of the geometrical magnitudes suffices for an argument that is not concerned with measuring, but only with relations between the magnitudes.

Thus, the concept appears to be closely connected with the theory of the X'th book; and we presume that the word *dynamis* developed a particular meaning when used of the objects of that theory, rectangular quadrangles. One of the most frequent operations in the X'th book is "squaring" a quadrangle: to any given rectangle there is one and only one square that equals the rectangle, i.e.: that has the same *dynamis* as the rectangle. But if the *dynamis* of a plane surface is conceived as "the extension of the square that has the same extension as the surface", - then it is no wonder that the word *dynamis* in some contexts seems to mean the same as *tetrágōnon*, "square". However, we understand the difference: whereas *tetrágōnon* means "rightangled equilateral quadrangle" (a geometrical magnitude), *dynamis* means the "extension" of such a quadrangle, its "capacity". As a matter of fact, we never meet with the phrase *tetragōnōi sýmmetroi*, "commensurable in square", but always with *dynámei sýmmetroi*, "commensurable in plane extension". To the author of the X'th book *tetrágōnon* and *dýnamis* were not synonymous.

3.53    More readers than one will ask: Who told you that *dynamis* means the "extension" or "capacity" of a plane surface? In the Elements and in Plato's dialogue *Theaitetos* 147e5 - 148b2 it appears to be used of line segments.    The answer is: Plato told us, in *Timaios* 31c4. Or to be more cautious: If *dynamis* meant "the extension of a plane surface", one might borrow it right away to denote "plane numbers" (i.e.: such that are products of two factors), particularly square numbers. And if *o(n)gkos* meant "extension of a spatial magnitude", one might borrow it right away to denote "spatial numbers" (i.e.: such that are products of three factors), particularly cubes. In *Timaios* 31c4 Plato uses the words *dynamis* and *o(n)gkos* in that way.

In the X'th book, however, the word *dynamis* is unmistakably used of line segments; we can easily see why: if the *dynamis*, "capacity", of a plane surface is identified with the *dynamis* of the equivalent square, then the *dynamis* is determined (*dedoménē*, given) as soon as the side of the square is determined. In that sense, the side of the square, though a line segment, may be said to "have the *dynamis*".

That some subject has a *dynamis* is to say that the subject *dýnatai*, "can" or "masters" something. Thus, if a line segment b has the *dynamis* of its square □ b, the Greeks say that b *dynatai* □ b.
And if □ b = a.c, then b *dynatai* the rectangle a.c.
From the lemma X 13/14 we learn explicitly that
if □ c = □ a + □ b, then c *dýnatai* a and b; and
if □ c - □ a = □ b, then c *dýnatai meízon*, "masters more", than a by the square on b.
The "over-mastering" of b over c (which on p. 42 we called "the square difference") is an important concept in the theorems of the X'th book.
From the XIII'th book we learn that if the square on a equals five squares on q (cf. p. 12), then a Greek would say that a *dynatai pentaplásion* of q, "a masters five times q".

3.54     There is hardly one single word that will do in English; we
acquiesce in following Heath and translate *dynámei sýmmetroi* by "commensurable in square", seeing that in fact it does mean that two squares
are commensurable. We might recommend the verb "masters" for the Greek
*dynatai*, as its Danish counterpart "magter" was very successfully used
by Thyra Eibe in her translation of the Elements into Danish (1897-1912).
Thus, e.g. X 57 would run in that idiom:

> If a quadrangle is contained by a red and the fourth sum of
> two reds, then the line segment that *masters* the quadrangle
> is obscure, the so-called Greater.

But this booklet being no translation, we will not bother.

Traditionally, there is some dispute whether *dynamis* means "square" or
"square root" (*), - an idle contention, if we are not mistaken. We
might be if we had to rely on Plato alone, who is anything but lucid in
*Theaitetos* and *Timaios*. We are satisfied that he does not contradict our
interpretation, which is corroborated by the doings in the Elements'
theory of quadrangles.

---

(*) Recent aperçus are Burnyeat, M. F.: "The Philosophical Sense of
Theaetetus' Mathematics", *ISIS* 69 (No. 249), pp. 489-513. 1978.
And Maurice Caveing, against A. Szabó, "Anfänge der griechischen Mathe-
matik", *Revue d'hisoire des sciences* XXXII, 2. April 1979.

## Bibliographical Epilogue

This book was not meant to be History of Science in the sense of, say, "The Evolution of the Euclidean Elements" by Wilbur Richard Knorr (Reidel, 1975). It was meant to be what it says: A Guide. To give access to the X'th book for those who may have lost faith in the commentairies by Thomas Heath, but who shrink from the long wanderings through the Greek text, - that they may partake in the growing debate about the nature of Greek (and pre-Greek) mathematics.

This may turn out to be *hybris*: after all, we ourselves were guided by nobody but the author of the X'th book of the Elements. Therefore, we shall guide nobody into the literature about the topic, but refer to Wilbur Knorr, who mentions every publication that can reasonably be said to bear on it.

If you have got the impression that the subject of the X'th book is a very limited one, that the reasonings are simple and straightforward manipulations with quadrangles in the idiom of commensurability, and that any association with theories of equations is *ex post facto* modernism (though not surprising, seeing that quadrangles are mathematical objects prone to measuring) - well, then we share an impression of that fine piece of literature. We had rather leave it at that.

København, 1982 March 7.

## Index of special terms

|  | page |
|---|---|
| amber line segment = *mésē*, amber quadrangle = *méson* | 26, 30, 67 |
| *álogos* = irrational = obscure | 26, 35 |
| apotome | 36, 62 |
| apply a quadrangle to a given line segment | 20, 30 |
| binomi(n)alis = *ek dýo onomátōn* = sum-of-two-reds | 36, 37, 62 |
| commensurable with B, A is | 16, 21 |
| complements = *paraplērōmata* | 19, 20 |
| contained by, any rectangle is | 19 |
| *dynamis* | 25, 42, 72 |
| *ek dýo onomátōn* = binomi(n)alis = sum-of-two-reds | |
| *elássōn* = Lesser Line (= Minor) | 9, 14, 49, 50, 53 |
| elliptical application of areas | 41, 42, 51 |
| equivalence relation, *com* is an | 24 |
| fits, the one that = *prosharmózousa* | 48 |
| golden section | 11, 19 |
| Greater Line = *Meízōn* (= Major) | 49, 50, 53 |
| lemmata, positions of | 17, 65 |
| Lesser Line = *elássōn* | |
| *mésos* = amber | |
| obscure = *álogos* | |
| quadrangles = squares & rectangles | 16 |
| red line segment = *rhētē*, red quadrangle = *rhētón* | 26, 27 |
| *rhētē rhētón* = red | |
| square difference, the = √(□ b - □ c) | 42, 43 |
| square only, commensurable in (cf. X def 2) | 25 |
| sum-of-two-reds = binomi(n)alis = *ek dýo onomátōn* | |
| uniqueness of splitting | 47, 56 |
| whole, the (= an apotome plus the one that fits) | 48 |